Machine Learning and Robotics in Urban Planning and Management

Kamalesh Ravesangar
Tunku Abdul Rahman University of Management and Technology, Malaysia

Christian Kaunert
Dublin City University, Ireland & University of South Wales, UK

Bhupinder Singh
Sharda University, India

Sahil Lal
Sharda University, India

Manmeet Kaur Arora
Sharda University, India

IGI Global
Scientific Publishing
Publishing Tomorrow's Research Today

Published in the United States of America by
IGI Global Scientific Publishing
701 East Chocolate Avenue
Hershey, PA, 17033, USA
Tel: 717-533-8845
Fax: 717-533-8661
Website: https://www.igi-global.com E-mail: cust@igi-global.com

Library of Congress Cataloging-in-Publication Data

Names: Ravesangar, Kamalesh, 1985- editor. | Kaunert, Christian, editor. | Singh, Bhupinder (Law teacher), editor. | Lal, Sahil, 1990- editor. | Arora, Manmeet Kaur, 1992- editor.
Title: Machine learning and robotics in urban planning and management / edited by: Kamalesh Ravesangar, Christian Kaunert, Bhupinder Singh, Sahil Lal, Manmeet Arora.
Description: Hershey PA : IGI Global Scientific Publishing, [2025] | Includes bibliographical references and index. | Summary: "There have been drastic changes in the field of urban planning and management with integration of advanced technologies such as machine learning (ML), autonomous vehicles etc, Given the importance of such innovations to modern urban traffic flow and congestion, it is imperative that they be protected"-- Provided by publisher.
Identifiers: LCCN 2024057477 (print) | LCCN 2024057478 (ebook) | ISBN 9798369394106 (hardcover) | ISBN 9798369394113 (paperback) | ISBN 9798369394120 (ebook)
Subjects: LCSH: City planning--Technological innovations. | Machine learning. | Robotics.
Classification: LCC HT166 .M2265 2025 (print) | LCC HT166 (ebook) | DDC 307.1/2160285631--dc23/eng/20250131
LC record available at https://lccn.loc.gov/2024057477
LC ebook record available at https://lccn.loc.gov/2024057478

British Cataloguing in Publication Data
A Cataloguing in Publication record for this book is available from the British Library.

Table of Contents

Detailed Table of Contents

Smart city management is being transformed through the incorporation of ML and robotics, which increase efficiency and sustainability as well as strengthen public services. Transformation is accompanied by vast ethical, legal, and policy challenges. ML-based applications for making decisions and autonomous robotic systems raise issues related to data privacy, security, and algorithmic bias that require a structured regulatory framework. The ethical dilemma comes from the balance between automation and human involvement, transparency, and disruption of socioeconomic inequalities. AI systems will allow a level of complexity and responsiveness that has not been possible until now and, indeed, will be a part of the smart urban infrastructure of the future. This chapter analyzes the multifaceted implications of ML and robotics on urban governance with clear policy recommendations to realize equitable, secure, and sustainable smart cities.

New-age marketing intelligence and AI are shaping the way the education sector harnesses consumer understanding in smart cities where digital infrastructure and technology integration is at its boom. AI-driven systems like predictive analytics and machine learning algorithms allow institutions to analyze large datasets,

planning and development as a whole with a closer look at potential safety concerns, issues of cyber security and the require for strong regulatory frameworks in the face of fast-paced change. It emphasizes the need for safety by design principles and robust verification practices to ascertain the reliability of AV systems. The chapter takes account of cybersecurity threats due to malware and data privacy issues stemming from mass data acquisition that is part of AV functionality. This explores different approaches to integrating AVs into the urban ecosystem while highlighting the importance of public-private partnerships and community involvement in building trust and acceptance.

Increased congestion, inefficiency, and accidents in cities are major issues for urban traffic systems. However, rapid urbanization and increasing numbers of cars exacerbate problems that have created an environment too dynamic and sophisticated for traditional solutions like static traffic signals or road expansion. The chapter discusses the use of machine learning and robotics with graph neural networks and reinforcement learning for optimizing traffic flow. Traffic networks pose intricate relationships that GNNs model under the form of nodes and edges representing roads, intersections, and vehicles. RL allows for continuous real-time interaction through which autonomous agents learn optimal strategies; thus, better decision-making takes place in dynamic traffic conditions and the system can proactively adjust signal timings, reroute vehicles, and manage congestion. Integration of these technologies will indeed be transformative to traffic management; hence, more effective, flexible, safest transportation systems will be expected in the future.

E-waste is a waste which is gathered from various sources, Household sector is considered as a biggest source of generation of e-waste. E-waste consists of various components characterized as hazardous and non-hazardous components also contain approximately 1000 substances categorized in this category. E-waste is comprising of ferrous and non-ferrous metals ceramics and other items. When e-waste gets dismantled and continuously processed it jeopardizes the health environment, and

Foreword

To tackle the urbanization challenge of the 21st century, new creative approaches must be developed to manage cities efficiently, sustainably and smartly. As global populations become more concentrated in cities, issues like traffic congestion, environmental degradation, housing shortages and infrastructure inefficiencies only grow. Emerging as transformative forces in its own right, ML and robotics empower urban development and management with data-driven decision-making, automation, and predictive analytics to improve the quality of urban life. This edited book on *Machine Learning and Robotics in Urban Planning and Management* is a comprehensive guide that brings together top-level experts and researchers to discuss the challenges and opportunities posed by these transformative technologies.

Machine learning equips urban planners with the ability to process large datasets on demographics, transportation patterns, land usage, and environmental factors. AI-based predictive models also help optimize traffic flow, bolster public transport systems and boost energy efficiency in buildings. On the other hand, robotics is transforming construction, waste management and disaster response through automation of complex processes and increased safety. Autonomous vehicles, drones, and AI-enhanced surveillance systems are collectively expanding urban security and emergency management. Though as cities incorporate these technologies, ethical questions regarding data privacy, algorithmic bias and equitable access need to be navigated thoughtfully to foster fair urban development.

I congratulate the editors Professor (Dr.) Bhupinder Singh, Sharda University India; Professor (Dr.) Christian Kaunert, Dublin City University Ireland & University of South Wales; Professor (Dr.) Kamalesh Ravesangar, Tunku Abdul Rahman University of Management & Technology Malaysia; Sahil Lal, Sharda University India and Manmeet Kaur Arora Sharda University India for this edited book which explores the diverse arena on urban management in the Age of AI, Robotics and Automation This book is a timely contribution that will be a source of ideas for policymakers, urban planners, and engineers and researchers who seek to integrate new tools of AI and robotics into the effort of managing cities. It offers a broad view of the potential of ML and robotics to shape future cities by looking at real-world use cases, policy frameworks, and technological advancements. With making this edited book, I hope to encourage intelligent and balanced approaches for realizing solutions in support of Smart Cities around the globe as we move into a more connected, automated and resource-conscious urban future.

Nishant Awana
Supreme Court of India, India

Preface

The edited book *Machine Learning and Robotics in Urban Planning and Management* comes at an important point in time, where cities across the globe are confronting unprecedented challenges. Urbanization, climate change and resource constraints are increasing the need for innovative solutions that can only be realized through the adoption of new technologies. The contributors to this book represent wide array of disciplines from machine learning, robotics, urban planning, and management. The book illustrates how diverse range of disciplines can come together to cross-pollinate, presenting cutting-edge thinking for urban development.

This compilation is inspired by the understanding that standard practices in the building or urban developments are not meeting in time required to face the challenges expected in the future cities. The introduction of machine learning and robotics offers novel opportunities to enhance decision-making processes, maximise resource allocation and improve the quality of life of people living in cities. Each chapter delves in to the applications of these technologies, including highlighting case studies and theoretical frameworks that illustrate their potential use. This book is intended to researchers, practitioners and policymakers interested in realizing the consequences of combining machine learning and robotics in urban environment planning. The book aspires to create an interdisciplinary conversation among academia, organizations and industries, motivating the reader to engage into deeper thought on how technology can help to form a perspective for sustainable urban futures.

CHAPTER OVERVIEW

This book offers a comprehensive exploration of how advanced technologies such as machine learning (ML) and robotics are transforming urban landscapes. Below is a chapter-by-chapter overview of the book:

Chapter 1: AI and Robotics in Smart City Governance: Ethical and Legal Pathways for Sustainable Urbanization

This chapter examines the role of AI and robotics in reshaping urban governance. It addresses the ethical, legal, and policy challenges that come with the integration of these technologies, such as data privacy, algorithmic bias, and the balance between automation and human involvement. The authors propose policy recommendations to guide the development of equitable, secure, and sustainable smart cities.

Chapter 2: Harnessing Marketing Intelligence and AI to Understand Consumer Behavior in the Education Sector in Smart Cities

In this chapter, the authors explore how AI-driven marketing intelligence is transforming the education sector within smart cities. By leveraging predictive analytics and machine learning, educational institutions can gain insights into consumer behavior, enabling them to tailor marketing strategies and deliver adaptive, data-driven education systems.

Chapter 3: Integrating Mobility in Multimodal Transportation Systems: A Comprehensive Analysis and Their Impact on Urban Mobility

This chapter focuses on the integration of various modes of transportation—cars, public transport, bicycles, walking, and micromobility—into multimodal transport systems. It addresses challenges such as coordination between providers, differing regulations, and the need for integrated ticketing solutions, offering policy recommendations for improving urban mobility and sustainability.

Chapter 4: Integration of Autonomous Traffic Systems for Urban Mobility and Urban Planning: Developing Policy Frameworks for Smart City

Autonomous vehicles (AVs) have the potential to significantly impact urban transportation. This chapter examines the implications of AV technology on urban planning, highlighting safety concerns, cybersecurity risks, and the need for regulatory frameworks. The authors advocate for safety-by-design principles and public-private partnerships to ensure the successful integration of AVs into urban ecosystems.

Chapter 5: Machine Learning and Robotics in Urban Traffic Flow Optimization With Graph Neural Networks and Reinforcement Learning

This chapter explores the application of machine learning, particularly Graph Neural Networks (GNNs) and Reinforcement Learning (RL), in optimizing urban traffic flow. The authors present how these technologies can be used to manage congestion, adjust signal timings, and reroute vehicles, leading to more efficient and flexible traffic management systems.

Chapter 6: Machine Learning in Reducing E-Waste: A Global Legal Perspective

The growing issue of electronic waste (e-waste) is addressed in this chapter, which discusses how machine learning models can help in the regulation and management of e-waste. The authors introduce predictive models for tracking e-waste generation and highlight the role of AI in categorizing and reducing harmful components of e-waste, offering a more sustainable approach to managing this growing concern.

Chapter 7: Revolutionizing Rural Development: Mobile-GIS and Cognitive Intelligence Systems as Case Study of Chhattisgarh

This chapter highlights the integration of AI with Mobile-GIS technology in rural development, using Chhattisgarh as a case study. The authors show how AI-driven cognitive systems are transforming sectors like agriculture, healthcare, and resource management, providing rural communities with tools for more sustainable and efficient development.

Chapter 8: Safety and Security in Autonomous Traffic Systems Management With AI: Economic Implications of Traffic Optimization in Smart Cities

Focusing on the safety and security aspects of autonomous traffic systems, this chapter explores the role of AI in optimizing traffic flow while ensuring robust cybersecurity measures. It examines the economic and environmental benefits of AI-driven traffic systems, including reduced congestion, lower fuel consumption, and enhanced sustainability.

Chapter 9: Smart Cities Ahead: Policy and Regulatory Strategies for Machine Learning and Robotics in Urban Development

This chapter investigates the policy and regulatory frameworks needed to support the adoption of ML and robotics in urban development. It surveys existing smart city initiatives, compares global approaches to governance, and examines how proper policy implementation can lead to improved resource allocation, service delivery, and citizen engagement in smart cities.

Chapter 10: Smart Parking Systems With Machine Learning: Working Mechanism on Environmental Impact and Traffic Optimization

This chapter focuses on the role of smart parking systems powered by machine learning in reducing traffic congestion. The authors discuss the environmental impact of urban traffic and propose solutions for optimizing parking, detecting accidents, and managing congestion, with the goal of improving urban traffic management and minimizing ecological harm.

Chapter 11: Sustainable Entrepreneurship in Urban Planning and Management: AI as an Enabler for Green Startups

This chapter explores the role of AI in enabling sustainable entrepreneurship within the context of green startups. It discusses how AI technologies can optimize operations, reduce carbon footprints, and foster innovation in environmentally conscious business models, empowering startups to thrive while contributing to sustainability goals.

Chapter 12: Sustainable Urban Mobility and Smart Traffic Management: Balancing Optimization With Environmental Sustainability Goals

The final chapter addresses the balance between urban traffic optimization and sustainability. By leveraging technologies like AI and IoT, the authors explore how smart traffic management can improve mobility, reduce greenhouse gas emissions, and promote sustainable urban transportation. The chapter emphasizes the importance of integrating traffic management with long-term environmental sustainability goals.

Each chapter in this book contributes a vital perspective on how machine learning and robotics are driving the future of urban planning, management, and development, with a focus on creating smarter, more sustainable, and equitable cities for the future.

We extend our heartful gratitude to everyone who contributed for their time and thoughts. This is a tremendous introduction to a broad area of pioneering research and as such will be invaluable in preparing researchers for the state of the art and, also, the path ahead. As our society continues to dance along the precipice of the digital age, let us embrace this opportunity to adapt and flourish, as we together explore the content presented on this platform in our common pursuit for smarter, more resilient cities.

Chapter 1
AI and Robotics in Smart City Governance:
Ethical and Legal Pathways for Sustainable Urbanization

Anjali Raghav
https://orcid.org/0009-0004-0248-7956
Sharda University, India

Bhupinder Singh
https://orcid.org/0009-0006-4779-2553
Sharda University, India

Richa Raghav
https://orcid.org/0009-0009-8482-6232
Sharda University, India

Kozma Dorottya Edina
https://orcid.org/0000-0002-4948-8815
Pannon Egyetem University of Pannonia, Hungary

ABSTRACT

Smart city management is being transformed through the incorporation of ML and robotics, which increase efficiency and sustainability as well as strengthen public services. Transformation is accompanied by vast ethical, legal, and policy challenges. ML-based applications for making decisions and autonomous robotic systems raise issues related to data privacy, security, and algorithmic bias that require a structured regulatory framework. The ethical dilemma comes from the balance between automation and human involvement, transparency, and disruption of socioeconomic inequalities. AI systems will allow a level of complexity and responsiveness that has

DOI: 10.4018/979-8-3693-9410-6.ch001

not been possible until now and, indeed, will be a part of the smart urban infrastructure of the future. This chapter analyzes the multifaceted implications of ML and robotics on urban governance with clear policy recommendations to realize equitable, secure, and sustainable smart cities.

INTRODUCTION

Smart Cities are needed in the modern world where the rise in population and restrictions on resources pave the way for urbanization at unprecedented speeds. Municipalities with a smart city definition embrace technology and use information and communication technologies (ICT) to optimize operations, communicate with residents, and even improve government services. The convergence of the Internet of Things, data analytics, cybersecurity, and advanced infrastructure is behind the transformation into smart digital cities, which use connected technologies that allow for real-time data collection and analysis to improve urban services. Smart city concept not only facilitates the capital growth but helps the sustainable cities development that meets the evolving urban metropolises needs (Satpathy et al., 2025). The importance of advanced technologies, specifically Artificial Intelligence (AI), Machine Learning (ML), and robotics, in shaping these one-day cities has become significant. Technology is an essential component of urban management today and in the future. Including IoT, AI algorithms, and data analytics to enhance monitoring and enrollment across a city, AI ensures traffic management systems optimize the traffic flow, minimize traffic, and increase road safety by analyzing live data. ML models can also predict usage patterns for various resources, such as electricity or gas, enabling better energy distribution and reduced waste. There are many areas in which cities can leverage these technologies, from addressing waste management to providing public safety and better environmental sustainability. Beyond its role of enhancing operational efficiency, the implementation of smart technologies also creates opportunities for citizen engagement as they offer transparent access to information and services (Padhiary et al., 2025).

Yet, the implementation of AI and technologies within the urban governance context carries with it a series of ethical, legal, and policy hurdles. Another significant issue is that cities are collecting enormous amounts of personal information from their citizens and, as a result, putting them at risk of data privacy breaches. It is essential to ensure that this data is managed responsibly and ethically while maintaining public trust. Lastly, there could also be algorithmic bias issues in AI systems that propagate existing social injustices if not countered appropriately. Also, the digital divide, if certain groups of the society don't have access to technology or they lack digital literacy skills, they risk being marginalized in the smart city narrative (Abbass et al., 2025). Global governance structures are needed to manage these changes and protect human rights in smart cities. Mainly, there is a requirement to investigate how Artificial Intelligence (AI) and Machine Learning (ML) can be incorporated into the foundation of urban management framework for improving the administration and process of services. This entails analyzing the current uses of these technologies across sectors including, but not limited to, transportation, healthcare, and public safety. Moreover, related to the previous point, studies should explore best practices to ensure responsible and ethical deployment of AI in urban areas dedicated to data governance and algorithmic accountability. After assessing trends in the applications of emerging technologies, it also reports on some of the challenges that municipalities are facing in the pursuit of smart city initiatives.

2. MACHINE LEARNING AND ROBOTICS IN SMART CITY DEVELOPMENT

Smart city management is defined as a broad move to integrate innovative and modern information and communication technologies (ICT) in urban governance for improving residents' quality of life. Data-driven decision-making is a key feature of smart city management that enables enhanced operational efficiency and delivery of public services. Smart city management involves the digitalization of multiple municipal functions, allowing real-time data collection and analysis for its efficient function. Such interconnectivity improves governance and supports sustainability, resulting in a more habitable urban experience for citizens (Hind et al., 2025).

2.1 Applications of Machine Learning in Urban Planning

Machine learning (ML) is an integral part of the urban planning process that allows cities to use data to make the best decisions. Traffic management and predictive insights are some of the leading use cases. ML algorithms can enable cities to ascertain traffic patterns, predict congestion, and optimize traffic signals accordingly. In addition, it optimizes traffic flow which helps in decreasing both travel time as well as emissions, thus promoting greener cities. For example, adaptive traffic signal systems that adjust at the moment according to the changing traffic conditions can be implemented by the cities to improve overall mobility. Energy efficiency and resource optimization is another very substantial area of Machine Learning applications. With machine learning capabilities, smart grids can analyze consumption patterns and anticipate energy demand fluctuations (Kaushik et al., 2024). This helps to optimize energy distribution and minimize wastage, in turn, minimizing the costs for both municipalities and citizens. For example, by using predictive maintenance models, cities can detect any possible failures of the infrastructure even before it is broken, thus limiting the inactivity and the expenses for the repair. Machine learning applications in waste management and sustainability. Cities can also deploy sensors to analyze rates of waste generation and optimize collection routes and schedules accordingly. Additionally, ML could facilitate recycling initiatives by detecting materials that are commonly disposed of together which would improve the sorting operations at recycling facilities. This aligns with sustainability objectives by minimizing landfill waste and encouraging recycling programs (Cities, 2025).

2.2 Importance of Robotics in Smart Cities

Smart Cities Application of Robotics Robotic process automation significantly improves the workings of smart cities. Autonomous public transport and mobility solutions are one such area. Self-driving vehicles (AVs) are being piloted and introduced in multiple cities across the globe to enhance public transport fleets. These vehicles can follow established routes or dynamically re-route based on actual demand, facilitating efficient transport and less capacity or congestion for the road. Robotics also plays an important part in surveillance and security systems that help maintain public safety. Flying cameras can help in monitoring the area to provide security or be used in emergency response to get a bird's eye view of the situation. Furthermore, robotic systems can also be integrated into surveillance networks to monitor and identify any suspicious activities or potential threats (such as robberies or terrorist attacks), allowing authorities to respond promptly, ultimately leading to a much safer urban environment. Another important application is the maintenance and automation of smart infrastructures (Lakshumiah et al., 2025).

Merits and challenges of AI-driven smart cities: AI-driven technologies for Smart City Initiatives Improved operational efficiency results in cost savings for municipalities and better service delivery for residents. City planners leverage data-driven insights to inform decision-making that enhances resource allocation and boosts overall quality of life. Additionally, AI technologies can enable enhanced citizen engagement through platforms that promote feedback and engagement in governance processes. Bringing about smart cities with AI, however, has its share of challenges. Security systems are still prone find be a major concern if the data, privacy can also be a major issue since some data will have to be leaked as the cities will collect huge amounts of personal information about its residents from several devices and sensors (Raghav et al., 2024). Properly addressing this data ethically and transparently is the key to maintaining public trust. Moreover, there are issues with algorithmic bias where AI systems trained from biased datasets can reinforce pre-existing disparities within an urban environment.

3. ETHICAL CONSIDERATION IN AI AND ROBOTICS IN DRIVEN URBAN MANAGEMENT

The accelerated adoption of technologies like artificial intelligence (AI) and robotics in urban management have revolutionized the functioning of cities, everything from transportation and security to energy efficiency and waste management. The use of these technologies has engendered ethical concerns even as they provide tremendous opportunities for improved quality of life. Artificial Intelligence Transformation Data Deployed in society the need for scrutiny of the extent to which privacy fairness and social equity are achieved (Ejaz et al., 2025). The key ethical issues in AI and robotics for urban governance: data privacy, algorithmic bias, socio-economic impacts, and human-centered AI are explored in this section.

3.1 Data Privacy and Surveillance Issues

Data Ethics in Smart Cities: Data privacy is one of the expected ethical challenges in AI-based urban management. The Internet of Things (IoT) has made cities smarter. These technologies gather and process large volumes of data, such as personal information, behavioral patterns, and real-time location tracking of residents. Data-driven insights based on modeling that could inform public services must, however, be balanced against questions of informed consent, security, and potential abuse of private information.

Robust governance frameworks that promote transparency, accountability, and explicit user consent underscore the ethical use of data. Data protection measures such as encryption, anonymization, and access controls need to be a priority for governments and tech companies to prevent unauthorized access or data breach. To build trust in AI-driven smart city solutions, residents should also know how their data is being captured, stored, and utilized.

Balancing Public Safety and Privacy Rights: For example, smart city infrastructures can use AI-powered facial recognition, predictive policing, and real-time monitoring systems to improve public safety. Although such technologies can facilitate crime prevention and emergency responses, they also pose risks to personal privacy and civil liberties. Data collection and analysis on such a large scale could ultimately lead to mass surveillance, and there are some concerns about government overreach and authoritarian control on the back end of AI for this reason. Finding the right balance between maintaining public safety and upholding the privacy rights of citizens calls for strong legal guidelines to determine how AI surveillance can be used and in what circumstances. The ethical application of AI in urban management must always adhere to three key principles: necessity, proportionality, and accountability. To that end, governments should create independent oversight bodies to review AI-driven surveillance programs and ensure that they are used responsibly, without curbing individual freedoms (Sathianarayanan et al., 2025).

3.2 Algorithmic Bias and Fairness

Discrimination in decision-making: AI is only as unfair as the data that trains it. Elsewhere, AI algorithms are being used in urban management, for things like law enforcement, hiring decisions, who gets social welfare, and who gets urban planning. At the same time, datasets biased for discrimination produce outcomes that discriminate, disproportionately affecting disadvantaged communities. For example, AI-based predictive policing has been criticized for perpetuating racial profiling and unfairly targeting certain socio-economic groups.

AI bias comes from past discrimination reflected in the training data, insufficient diversity in the algorithm development team, and imperfect decision-making models. If not addressed now, AI-driven urban governance could accelerate existing systemic discrimination rather than reduce it, exacerbating social inequalities (Das et al., 2025).

Mitigating Bias Through Ethical AI Development-To promote fair and unbiased AI systems is essential to integrate ethical AI design principles which are as follows:

1. Ensuring that then
2. datasets are diverse and representative is crucial in preventing algorithmic discrimination.
3. Performing regular bias audits and transparency studies of AI models.
4. Deployment of explainable AI (XAI) to increase interpretability and accountability of AI decisions

Governments and developers must come together with ethicists, social scientists, and affected communities to develop AI systems based on human rights and social justice values. Moreover, public participation in AI governance will also help expose biases and hold institutions accountable for their use of AI (Mahto et al., 2025).

3.3 Social and Economic Impacts

Unemployment and Job Automation: AI and robotics working in a smart city have also caused concern over potential labor displacement and job automation. These AI-based systems are being embraced increasingly for tasks like administration, security, and service, leading to a dwindling demand for human manpower in public transport systems, customer service, and municipal processes. On one hand, AI increases efficiency and productivity in various sectors, yet it disrupts conventional employment paradigms, leaving millions vulnerable to job displacement and contributing to growing income inequality.

For AI-driven urban management, one of the ethical responsibilities will be to manage the transition for those workers impacted by automation fairly. Governments can invest in reskilling programs, promote digital literacy, and create new work opportunities in AI related industries.

Ethical AI Implementation for Social Equity: Urban AI solutions have a responsibility go beyond just optimization and ensure social inclusiveness and equity. Without equitable distribution of the benefits resulting from AI and robotics, wealth disparities between socio-economic groups will continue to grow. Police must also ensure that AI-powered services are available to all citizens, including low-income communities, old-age people, and people with disabilities.

Smart city technologies should be designed and implemented according to universal design principles and bring equal access to public services, healthcare, transportation, and digital infrastructure. It is also essential that public officials and government agencies interact with local communities to know their specific needs and concerns because AI-powered urban strategies have to help the citizens and be citizen-centric (Suryawanshi et al., 2025).

4. LEGAL AND REGULATORY FRAMEWORKS FOR SMART CITY WITH AI

The emergence and rapid deployment of artificial intelligence (AI) and robotics in smart city governance highlights the urgent need for legal and regulatory frameworks that address the challenges of privacy concerns, accountability, intellectual property (IP) rights, and protection of data. AI systems may enable new levels of efficiency in cities, but their use in public decision-making poses critical legal and ethical challenges. Governments around the world are striving for holistic models of AI governance that can enable the responsible rollout of the technology while enabling innovation. This part analyzes extant laws and regulations, as well as intellectual property and data protection laws and legal issues that inform AI-informed urban governance.

Existing Legislation on AI and Robotics: International Legal Framework

The legal framework for AI and robotics in smart cities is still being developed, with international institutions and national governments working towards the implementation of standard rules. The OECD AI Principles, adopted across 46 countries, promote the development of AI systems that are transparent, accountable, and human-centered. Analogous to this, the European Commission's AI Act organizes AI applications by risk levels and thus places severe regulatory requirements on high-risk AI systems including, but not limited to AI systems used in law enforcement, biometric surveillance, and critical urban infrastructures (Chavda et al., 2025).

UN and WEF promote international collaboration in AI governance, calling for legal frameworks that transcend borders to ensure standardized deployment of AI around the world. The IEEE Global Initiative on Ethics of Autonomous and Intelligent Systems also offers guidelines for responsibly developing AI, including factors like data protection, compliance with human rights, and ensuring AI safety.

Country-Specific Smart City AI Regulation – Many countries have implemented country-specific regulations to regulate the use of general AI applications in urban settings:

European Union (EU): The AI Act and General Data Protection Regulation (GDPR) provide strict regulations on AI transparency, bias prevention, and the protection of citizens‘ data.

United States: The US does not have a comprehensive regulatory framework for AI, and existing US AI regulations are limited to specific sectors, with fairness and accountability in AI decision-making featuring heavily in initiatives like the Algorithmic Accountability Act and the AI Risk Management Framework (Dutta et al., 2023).

China: The government implements a rigid AI governance policy, particularly in the area of facial recognition and smart surveillance, to preserve socal order. The Personal Information Protection Law (PIPL), is in line with global data protection practices.

India: National Strategy for AI (NSAI): The NSAI aims to decimate the 11 pillars of the national AI agenda for ethical, inclusive, and responsible development and deployment of disruptive technologies in urban infrastructure.

United Kingdom: The AI White paper supports a pro-innovation approach, taking a two-pronged approach towards regulatory oversight balanced with flexibility for AI developers

Legal Challenges in AI Empowered City Management

Liability and Accountability in AI Decision-Making: Arguably the most urgent legal challenge in AI-enabled urban governance is the issue of liability. For example, it is uncertain who is legally liable when an AI-powered traffic management system causes an accident due to faulty algorithms. Some jurisdictions have called for strict liability frameworks wherein AI developers or system operators are held directly accountable for damages caused by their algorithms. Others suggest shared liability models, where responsibility is spread across AI manufacturers, data suppliers, and policymakers (Elsayed & Mohamed, 2025).

Current AI liability laws lack homogeneity, therefore, enforcement becomes challenging, which reveals the need for a context-specific legal framework that precisely elaborates risk allocation and compensation pathways through urban systems driven by AI (Lal et al., 2024).

Regulation of Autonomous Systems in Urban Spaces

Smart city development may involve using the Internet of Things for autonomous vehicles, healthcare drones, and robotic public services, along with the need for rules on accountability, safety, ethics, lawfulness, references to pilot projects, and public interest (Hammouch & Raghav, 2025). Many cities have regulatory gaps regarding:

- Self-driving cars powered by AI: Governments will need to create standardized processes for licensing, insurance, and safety assessments.
- Drone surveillance: The regulations should include a solution for privacy violations, airspace management, and real-time data collection ethics.
- AI for Law Enforcement: It would be unwise to let AI-powered predictive policing and facial recognition risk being misused without legal oversight.

For example, the UK and Germany have implemented very robust AI safety certification requirements for autonomous systems, but standardized measures are still not globally recognized.

Artificial Intelligence Innovations and Intellectual Property

Ownership of AI-Generated Urban Solutions- With the increasing prevalence of AI in urban planning, IP laws must evolve to delineate ownership rights regarding AI-generated innovations. Some applications of AI-powered algorithms in smart city management are:

- Urban traffic control predictive analytics
- Artificial Intelligence-generated plans for infrastructure development
- Machine learning algorithms to minimize energy usage

Another legal quandary is if AI spontaneously creates a new urban design solution, who owns the IP, the person who created the AI, the municipality, or the AI itself? Existing IP frameworks do not recognize AI as an inventor, meaning that human developers or organizations usually receive credit (Xu & Guo, 2025). But experts disagree about whether AI should retain co-inventorship rights which would necessitate amendments to patent laws and copyright regulations.

Open Data Policies and Proprietary AI Models

These topics of artificial intelligence, data ownership, and proprietary algorithms are another area of contention in the governance of smart cities. Read More: How AI Will Reshape Your CityLike open data policies, which are meant to encourage transparency and innovation, tech companies frequently keep AI-laden solutions within cities proprietary through the use of a licensing model (Sepehri et al., 2025). This leads to legal tensions regarding:

- Open AI-powered urban analytics
- Algorithms of decision-making in smart cities under corporate control
- AI Fairness with Trade Secret Obligations

Regulatory frameworks must balance protecting the public interest (relevant stakeholders) while incentivizing private-sector advancements and minimizing monopolistic tendencies; thus, there is a need to ensure AI innovations are benefiting society rather than cementing monopolies.

Adherence to Data Protection Laws

GDPR, CCPA, and Other Global Privacy Laws

However, AI-based smart cities would involve the processing of large-scale personal and sensitive information, which would mean strict compliance, not just with local laws, but also with global data protection laws (Lubis et al., 2025). Key regulations include:

- General Data Protection Regulation. (GDPR). (EU): Requires consent from citizens to be processed by AI, requires minimization of data, and requires the right to explanation for decisions made by AI.
- California Consumer Privacy Act (CCPA) (USA): Enables citizens to opt out of AI-processed data, and to request for the data to be deleted.
- Personal Data Protection Act (PDPA) (Singapore): Provides rules regarding AI data security and legal sharing of data.
- China PIPL: Tight controls over cross-border data flow, impacting the operation of global AI partnerships.

In the context of AI implementation in smart cities, compliance with the laws is imperative in precluding mass surveillance, data breaches, and misuse of personal information.

Ethical Handling of Citizen Data

From a legal standpoint, the AI governance framework must not only comply with laws but also maintain ethical standards in handling citizen data. Governments should:

- Make public AI's role in collecting urban data for accountability.
- Use privacy-preserving techniques such as federated learning and differential privacy.
- Require independent audits on the effects of AI on privacy and human rights.
- Human-centric AI governance will help orient smart city innovation toward public trust, fairness, and ethical responsibility.

5. POLICY IMPLICATION AND RECOMMENDATIONS FOR AI IN SMART CITIES

AI and machine Learning in smart cities: Artificial Intelligence is a key aspect of the smart city context. Advanced AI technologies like smart traffic systems, predictive policing, and automatic energy grids could change the cities. Nonetheless, there are serious challenges that come with these advances related to ethics, privacy, security, and regulation. These challenges require proactive strategies from policymakers to ensure that AI and ML are governed responsibly and ethically across sectors (Nechesov et al., 2025). Well-designed governance of AI systems is crucial to balancing the public interest and minimizing bias, surveillance, and job displacement risks.

Policies to Ante-up on Proactive AI Governance: Constantly updated AI policies are required, as it rapidly develops in today`s climate. Smart city applications of AI are dynamic and fluid, and traditional regulatory frameworks tend to be static and poorly suited to such a pace of change. At the same time, policymakers need to create flexible oversight that adapts as AI advances (Arora et al., 2024). These include the understanding that regulations must be agnostic to the species of AI that emerges, as technical developments will always outpace rules, and the need for rules to apply to services powered by different AI capabilities. Following a myriad of stakeholders from urban planners to AI specialists, lawyers to citizens is necessary to develop public policies that serve everybody. Moreover, the concept of regulatory sandboxes safe environments for developing and testing AI applications can provide policymakers with an important tool to evaluate potential pitfalls before full-scale implementation.

Policymakers are instrumental in determining the pace of AI acceptance in smart cities. They have to work to ensure that AI-driven systems respect human rights, fairness, and transparency. You can do this by establishing accountability for AI decision-making, and by requiring transparency and audit requirements for AI models. Governments need to promote the development of ethical AI practices and facilitate public-private partnerships to encourage the development of responsible AI applications. Moreover, recognizing that job displacement will be a real concern, investment will also need to be made into AI education and retraining programs as AI-driven economic transformations take place, to ensure that societies are equipped to adapt to the changing job landscape.

Ensuring Ethical Deployment of AI: The role of AI in smart cities cannot be overlooked, but the use of AI must be undertaken transparently, inclusively, and ethically to ensure successful integration. The participation of the general public in the governance of artificial intelligence is an integral part of this process. However, AI-powered urban management without meaningful citizen engagement is prone to becoming viscous and not accountable. To combat this, governments should implement open AI governance platforms that allow residents to see applications of AI Already in use, review AI assessments, and submit feedback. Public consultations, AI town halls, and citizen advisory panels to debate and understand the implications of AI in the city's infrastructure (Suleimany et al., 2025). AI literacy programs also will be critical to helping educate the public about the benefits and risks of AI technologies, encouraging a well-informed society that can participate in AI policy debates.

Regulating AI-Powered Public Services: AI is increasingly deployed to enhance urban services including transportation, security, and energy management. Yet, the key to discouraging discrimination and eroding civil liberties hinges on fairness, transparency, and accountability in these domains. In transportation, where AI-fueled traffic management systems and self-driving cars can pose safety hazards if not adequately regulated, regulations are needed to produce equitable access to mobility solutions. This calls for transparent AI-driven security measures to be taken so that misuse does not occur in the form of biased facial recognition and mass surveillance, which could violate the rights of citizens. Moreover, guidelines for AI-based energy management policies must also mitigate efficiency challenges and ensure environmental sustainability while promoting clean energy solutions for all.

Policymakers must create a well-defined set of guardrails for AI applications in these industries AI-fueled public services must be regularly audited and come under public scrutiny to hold service providers accountable. Such new institutions, be they independent regulatory bodies to oversee the influence of AI on public services, can help minimize unintended consequences, such as algorithmic bias or discriminatory enforcement of AI-driven policies (Al-Rimawi & Nadler, 2025).

Cross-Sector Collaboration and International Cooperation

Partnerships among governments, technology firms, academia, and civil society organizations are vital for the development, deployment, and governance of AI in smart cities. Keeping in mind that responsible deployment of AI is core to its timely realization, Public-private partnerships (PPPs) overlap to ensure the benefits from AI in enhancing the utilization of technological know-how and monetary sources from the non-public sector. These collaborations must detail how they will respect data privacy, security, and accountability along the way. Supporting open-source AI projects and knowledge-sharing initiatives can also promote inclusive and accountable AI innovation (Gupta et al., 2025). Due to the global nature of AI governance challenges, international cooperation is necessary to develop harmonized policies and best practices. AI-enabled urban management has repercussions on global problems such as climate change, cybersecurity, and digital inequality, and governments must inevitably cooperate to develop harmonized AI regulatory standards. Ethical frameworks for international AI governance models should thus be developed that prevent AI abuse and promote responsible AI development. Global data-sharing frameworks can also improve urban planning and policy coordination to guarantee that AI solutions are used to the benefit of societies globally.

6. CASE STUDIES OF AI AND ROBOTICS IN SMART CITIES

AI and robotics have reshaped cities all around the world in urban management, transportation, governance, security, and citizen engagement. Many AI-based smart solutions have emerged from a handful of cutting-edge cities, establishing models for what the future may hold. After reviewing four major smart city case studies (Singapore, Barcelona, Dubai, and Amsterdam), exploring key ways in which AI can and is being leveraged to improve various aspects of smart city management.

6.1 Singapore: Artificial Intelligence (AI) in Urban Mobility and Sustainability

Singapore has been a global pioneer in harnessing AI for urban mobility and sustainability. AI-powered solutions in public transport, traffic management, and environmental preservation have been adopted by the city-state." Some of the most pioneering projects they are running include the Intelligent Transport System (ITS), which uses AI and machine learning to predict congestion on the road ahead, optimize traffic lights, and give commuters real-time data. This includes integrating artificial intelligence (AI) powered autonomous vehicle trials, which could pave the way for self-driving taxis and public transport options across future LTA developments. Singapore's Smart Nation Initiative integrates AI into energy management and environmental sustainability. AI algorithms are used to analyze energy usage patterns in both residential and commercial buildings, enabling owners to operate on optimal energy use and lower carbon emissions (Singh et al., 2025). The use of AI-enabled drones and robotics in the waste management sector and water conservation has immensely enhanced the efficiency of urban sustainability programs.

6.2 Smart Governance Driven by Big Data: Barcelona

Barcelona has become a world leader in data-driven governance, integrating AI and robotics into public services, citizen engagement, and municipal institutions. The Sentilo platform is an AI-driven urban data infrastructure that collects and aggregates real-time data from several IoT sensors installed throughout the city. Air quality, noise pollution, traffic patterns, and waste management are just a few examples of what these sensors monitor, which gives city officials insights that inform critical policy decisions (Gokulapriyan et al., 2025). AI-based chatbots and virtual assistants here are also major actors in citizen engagement and public administration. Barcelona's AI-fueled citizen services portal allows people to obtain government services, report urban problems, and receive live updates on city happenings. AI algorithms assist in the distribution of resources to make it more efficient and ensure that public services like healthcare, emergency response, and education are delivered to the individuals who require them the most.

One of the most impactful examples of AI-enhanced smart governance in Barcelona is the Decidim Barcelona initiative, a participatory democracy platform that enables citizens to input on urban development decisions. Using AI analytics, city planners take the public's input and integrate community preferences into policymaking. Barcelona's work around AI and transparent, data-driven decision-making demonstrates how smart governance can be used to achieve more efficient, responsive, and trusted government by citizens.

6.3 AI-Powered Surveillance and Security in Dubai

Dubai has also established itself as a global center for AI-based surveillance and security, embedding cutting-edge AI and robotics into public safety and law enforcement. Using analytical video spent on the AI-controlled surveillance system of the Dubai police to improve crime prevention and security observation. Based on Artificial Intelligence, the Oyoon (Eyes) project acts as a surveillance network that uses machine learning and computer vision to monitor potential threats, identify activities of suspicion, and support law enforcement agencies in real-time (Kumar et al., 2025).

Alongside surveillance, robotic police, powered in part by AI, have taken a walk in Dubai. Equipped with natural language processing (NLP) algorithms, these robots communicate with citizens, assist them, and process reports. One particularly innovative approach to securing the city is AI-powered predictive policing, where AI uses historical crime data to predict and prevent criminal activities as early as possible. AI is also important in disaster management and emergency response in the Emirate, apart from law enforcement. AI-powered traffic monitoring systems control the movement of vehicles during emergencies, thus making it possible for ambulances and firefighters to reach more quickly (Raghav et al., 2024). It has also been using autonomous drones during major public events to monitor crowds and for surveillance. This example from Dubai of AI in security shows how tech can be employed to maintain safety and justice. But it also poses important questions about privacy, the ethics of surveillance, and the balance between security and civil liberties.

Ethics of AI and Citizen Involvement in Amsterdam

Amsterdam has adopted ethical AI and citizen participation as foundational elements of its smart city programs. It is reputed to transparent AI governance and responsible data usage, thus public applications of AI in the urban environment have to conform to human rights and social values. An AI Project that builds ethical AI models for public services is Fair AI Lab, one of the largest AI influencers in Amsterdam. AI applications are held to standards of fairness, lack of bias, and accountability through collaboration with academic institutions and civil society organizations, as well as oversight from the city government. Regular audits are carried out for housing allocation, social welfare, and urban mobility to avoid discrimination and algorithmic biases in AI-based decision-making. In Amsterdam, initiatives such as AI for Good engage citizens via discussions about AI ethics and policy. Data transparency portals enable cities to share information related to AI usage in city management with their citizens. Called open governance, Amsterdam guarantees that AI will be a tool for the public good, not for unchecked surveillance or corporate hegemony. Additionally, the city of Amsterdam has implemented AI-driven sustainability initiatives, including intelligent energy networks that maximize energy consumption and navigational AI in boats for optimal travel across water bodies. These innovations underscore Amsterdam's focus on ethical integration, environmentally conscious development, and equitable urban design.

7. CHALLENGES AND FUTURE DIRECTIONS

While the advantages of AI robotics in smart city management are plenty, there are significant challenges that come with these innovations as well. From technological roadblocks to legal and ethical dilemmas, cities must navigate these complexities to ensure that AI is deployed in sustainable and responsible ways. This section discusses major issues in the way of technology integration, cybersecurity threats, legal and policy voids, and future perspectives on AI-based governance in smart cities.

Technological Challenges

One of the major obstacles to the integration of AI in smart cities is integrating AI systems into existing urban infrastructure. Many cities are still dependent on infrastructure that is only traditional and does not have the digital connectivity required for AI applications. Adding AI-enhanced solutions to already-existing systems requires considerable investments in hardware, software, and connectivity. Additionally, interoperability threats with multiple AI-assisted systems, including transportation management, energy grids, and public safety networks, pose another major challenge. Common data access, sharing, and synchronization are what AI-driven single SLOPs require, yet fragmentation and division across various departments and stakeholders make sharing this myriads of data types complicated (Vashishth et al., 2025).

One of the other tech threats is cybersecurity threats to smart city networking. Smart cities that are powered by AI, depend widely on huge amounts of real data that gets simulated from IoT sensors, cloud computing, and machine learning algorithms. This web of systems is rife with potential attack vectors for cyberattacks, data breaches, and AI subversion. These vulnerabilities can be exploited by malicious actors to disrupt transportation systems, compromise surveillance networks, or manipulate public services. Next-generation smart cities will rely on the integration of AI technologies coupled with advanced cybersecurity solutions to secure these networks and protect data privacy. But as cyber threats continue to evolve at a rapid pace, city administrators and cybersecurity experts have an ongoing battle on their hands.

Legal and Policy Gaps

However, the legal and policy landscape for AI within smart cities is still underdeveloped. As a result, existing laws and regulations are out of sync with the fast-moving development of AI. Many existing legal schemes around the world that are meant to govern, for instance, privacy, liability, and accountability were designed for more traditional governance models and therefore do not necessarily take into account the complexities of autonomous decision-making, data ownership, or AI-enabled surveillance.

A significant legal challenge has been in adapting existing frameworks to AI advancements. The traditional regulatory techniques may not be enough to address the algorithmic bias, automate decision-making, and the liability of AI. For example, AI-empowered law enforcement and predictive policing challenge civil rights, discrimination, and due process. In the absence of definitive legal frameworks, such AI models could exacerbate existing inequities in society. AI-specific policy, governance, and regulation should be added to existing structures in the form of new legislation that establishes equitable, transparent, and accountable processes while ensuring human agency. The data used for training extends only until October 2023. AI technology progresses rapidly, and state regulatory frameworks may quickly become out-of-date — even over a handful of years. However, to have a full, effective, and positive implementation, governments should devise agile and cross-evaluated policies to ensure monitoring, evaluation, and adjustment in parallel with continuous real-functioning practices of AI implementations. To establish AI regulations that are agile and proactive to strike a balance between innovation and ethical considerations, a multi-stakeholder approach that brings policymakers, technologists, legal experts, and citizens on the same page is essential (Sowjanya et al., 2025).

Future Trends in AI-Driven Urban Governance

AI-enabled urban projects are anticipated to see quite a lot of progress. A few emerging trends will likely define the future of AI-powered urban governance. One of the most promising use cases is AI-powered disaster management. As climate change and urbanization increase the likelihood of natural disasters, and because AI is capable of performing better at prediction, monitoring, and response capabilities. AI-based early warning systems analyze real-time data from weather sensors, satellite imagery, and historical disaster patterns to make accurate predictions of hurricanes, floods, and earthquakes. AI-enabled drones can also be used for search and rescue missions, and AI-supportive communication platforms can deliver emergency alerts and evacuation orders in real-time. These technologies will be key to reducing the loss of life, optimizing the response to disasters, and enhancing urban resilience.

Adopting Ethical AI in Urban Planning and autonomous Governance is also another prominent trend. The smart city of the future will include AI infrastructure that plans urban design models for optimizing land use, transportation networks, and environmental sustainability. Urban Planning AI can study population growth, traffic flows, and energy consumption trends to recommend urban development strategies that avoid congestion, pollution, resource depletion, etc. However ethical AI capable of addressing these urban ecosystem complexities will only be achieved through robust governance mechanisms, transparency of AI decisions, and active citizen engagement.

Moreover, AI-enabled workflows for the drive in governance are estimated to enhance efficiency in the delivery of public services. Adopting AI-powered digital assistants and chatbots will streamline and simplify interactions between government and citizens, allowing residents to quickly access needed permits, report issues, and receive real-time information. AI will also support policymakers in data-driven decision-making, as it will enhance efficiency in collaborative efforts, the allocation of the budget, social welfare, the management of resources for law enforcement, etc. However, the imperative of AI to be responsible, ethical, and in line with democratic values will need to be upheld through continuous transparency, governance, and digital ethics audits (Saini et al., 2025).

8. CONCLUSION

Artificial intelligence (AI) and robotics being integrated into smart cities have revolutionized urban management for effective governance, sustainability, and service delivery. That being said, figuring out how to regulate and govern the use of AI-powered systems poses difficult ethical, legal, and policy challenges. Key findings from the study, the need for the ethical design of governments' AI policies and regulations, and recommendations for more sustainable AI policies are given in this section. Artificial intelligence (AI) and robotics are leading to smart solutions for urban management that allow for streamlined transportation, security, and resource management. In case studies across Singapore, Barcelona, Dubai, and Amsterdam, how AI-driven solutions are making cities greener, smarter, and more urban mobile. AI implementations in public safety, energy management, and environmental monitoring have demonstrated promising prospects in promoting urban resilience. While new AI applications can bring significant benefits, they require a substantial amount of investment in existing infrastructure, and interoperability between many different AI-driven systems is far from being achieved. Moreover, challenges related to cybersecurity threats, algorithmic biases, and lack of regulatory frameworks continue to threaten the integrity and merit of equitable and secure AI implementation. The legal framework around AI in smart cities is still developing, with some data protection, responsibility, and AI ethics governance gaps. The absence of appropriate regulatory frameworks could result in all types of problematic behavior about the use of AI in urban environments, from breaches of privacy and discrimination to decreased human oversight. In essence, one of the main obstacles to AI-based urban governance is finding a balance between technological advancement and ethical as well as legal aspects. Though AI does empower cities to be more efficient and intelligent, it should never come at the expense of human rights, data privacy, and social equity.

Transparent decision-making, accountability mechanisms, and citizen participation are key to ensuring ethical deployment. AI-based surveillance and law enforcement frameworks, for example, need stringent regulation to avoid oversteps in government control, surveillance misuse, and unfair algorithms. AI ethics committees within smart cities can be established to oversee AI use in public governance. Holding discussions with people other than experts is important so that AI solutions belong to society both in terms of values and human rights, for instance. On the legal front, policymakers need to create adaptive legal frameworks that evolve along with AI developments. Laws related to urban planning, privacy, and liability, among others, are often antiquated when it comes to AI-driven smart city environments. Governments must establish agile, technology-specific legal regimes that cover areas like AI accountability, the prevention of bias, and resistance to cyber-attacks. AI and robotics have a unique potential to reshape urban governance, resulting in cities that are more durable, efficient, and adaptive. However, harnessing this potential will require a balanced approach integrating innovation with strong ethical and legal protections. As cities around the globe work towards building better cities using AI, issues like cyber threats, legal uncertainties, and ethical dilemmas need to be actively addressed.

Vision of the Future of AI for Smart Cities: Must underscores the need for responsible AI governance and dynamic policy-making Cross-sector collaboration, citizen engagement, and ethical AI frameworks for development are key areas for governments to implement in adopting AI to maximize its potential while establishing fair, secure, and inclusive uses.

REFERENCES

Abbass, M., Abbas, U., Jafri, R., Arif, S. M., & Akhai, S. (2025). AI and Machine Learning Applications in Sustainable Smart Cities. In *Sustainable Smart Cities and the Future of Urban Development* (pp. 1–32). IGI Global Scientific Publishing.

Al-Rimawi, T., & Nadler, M. (2025). Leveraging Smart City Technologies for Enhanced Real Estate Development: An Integrative Review. *Smart Cities*, 8(1), 10. DOI: 10.3390/smartcities8010010

Arora, M. K., Lal, S., Raghav, A., & Mishra, S. (2024). The Legal Implications of Corporate Borrowing and Analysing the Framework for Debt Financing and its Impact on Real Estate Sector in India. *National Journal of Real Estate Law*, 7(2), 1–4.

Chavda, B., Pillai, R., & Srivastava, A. (2025). Sustainable Migration: Unraveling the Connections Between Urban Planning, Migration, and AI. In *AI Strategies for Social Entrepreneurship and Sustainable Economic Development* (pp. 47-68). IGI Global Scientific Publishing.

Cities, S. (2025). 7 AI-Optimized Urban Resource Management for Sustainable. *Financial Inclusion and Sustainable Development in Sub-Saharan Africa*, 96.

Das, D., Bapat, J., Katsenou, A., & Shrestha, S. (2025). Connectivity for AI enabled cities—A field survey based study of emerging economies. *arXiv preprint arXiv:2501.09479.*

Dutta, P. K., Singh, B., Lal, S., Arora, M. K., Raghav, A., Mensah, G. B., & Alkattan, H. (2023). Encoding IoT for patient monitoring and smart healthcare: Connected healthcare system fostering health 6.0. *Babylonian Journal of Internet of Things*, 2023, 48–58. DOI: 10.58496/BJIoT/2023/007

Ejaz, U., Ramon, W., & Olaoye, G. (2025). The Role of Big Data and AI in Smart Cities and Urban Planning.

Elsayed, A., & Mohamed, M. (2025). Enhancing Smart City Management with AI: Analyzing Key Criteria and their Interrelationships using DEMATEL under Neutrosophic Numbers and MABAC for Optimal Development. *Neutrosophic Systems with Applications*, 25(2), 1–38. DOI: 10.61356/j.nswa.2025.25480

Gokulapriyan, R., Vinod, L., Chandran, R. A., & Yedla, V. R. (2025). Smart City Projects Influencing Smart Tourist Destinations. In *Advancing Smart Tourism Through Analytics* (pp. 333–350). IGI Global Scientific Publishing.

Gupta, S., Dwivedi, A., & Kumar, H. (2025). Digital Platforms as Catalysts of Circular Economy in Pursuit of SDG 11: A Rigorous Exegesis of Smart Cities. In *Innovating Sustainability Through Digital Circular Economy* (pp. 373-390). IGI Global Scientific Publishing.

Hammouch, H., & Raghav, A. (2025). AI and ML Applications in Wildlife Conservation and Forest Management: A Comprehensive Review. *Machine Learning and Internet of Things in Fire Ecology*, 49-68.

Hind, H., Lal, S., & Arora, M. K. (2025). Artificial Intelligence and Urban Sustainability. *Developing Smart Sustainability With Artificial Intelligence*, 17.

Kaushik, T. K., Singh, A., Raghav, A., Singh, B., Arora, M. K., & Lal, S. (2024). Untapping Artificial Intelligence in Environmental and Sustainable Progression: Projecting Legal and Ethical Considerations. In *Maintaining a Sustainable World in the Nexus of Environmental Science and AI* (pp. 109-130). IGI Global.

Kumar, A. A., Suryadevara, R., Sowmyya, T., & Chanal, G. B. (2025). Exploring the Intersection of AI, Healthcare, and Environmental Sustainability: Future Trends and Challenges. *Transforming Healthcare Sector Through Artificial Intelligence and Environmental Sustainability*, 23-47.

Lakshumiah, A., Malaiarasan, A., Packianathan, R., Natarajan, S. K., & Arumugam, G. (2025). Application of Artificial Intelligence (AI) Techniques for Green Transportation in Smart City. In *Driving Green Transportation System Through Artificial Intelligence and Automation: Approaches, Technologies and Applications* (pp. 335–358). Springer Nature Switzerland. DOI: 10.1007/978-3-031-72617-0_18

Lal, S., Raghav, A., Arora, M. K., Mishra, S., & Singh, B. (2024). Impact of Artificial Intelligence in the Health Sector: Potential and Barriers. In *Intersections of Law and Computational Intelligence in Health Governance* (pp. 100–116). IGI Global. DOI: 10.4018/979-8-3693-5976-1.ch005

Lubis, S., Nurmandi, A., Ahmad, J., Purnomo, E. P., Purwaningsih, T., & Jovita-Olvez, H. D. (2025). AI and Blockchain For Next-Generation E-Governance: A Comprehensive Bibliometric Review In Smart City Innovation.

Mahto, M. K., Srivastava, D. K., & Kaswan, M. S. (2025). Applications: Smart City and Urbanization Systems. In *5G Enabled Technology for Smart City and Urbanization System* (pp. 194-219). Chapman and Hall/CRC.

Nechesov, A., Dorokhov, I., & Ruponen, J. (2025). Virtual Cities: From Digital Twins to Autonomous AI Societies. *IEEE Access : Practical Innovations, Open Solutions*, 13, 13866–13903. DOI: 10.1109/ACCESS.2025.3531222

Padhiary, M., Roy, P., & Roy, D. (2025). The Future of Urban Connectivity: AI and IoT in Smart Cities. In *Sustainable Smart Cities and the Future of Urban Development* (pp. 33–66). IGI Global Scientific Publishing.

Raghav, A., Singh, B., & Jermsittiparsert, K. (2024). Role of artificial intelligence and machine learning in clean and blue energy for enhancing efficiency and sustainability. In *AI Applications for Clean Energy and Sustainability* (pp. 366–386). IGI Global. DOI: 10.4018/979-8-3693-6567-0.ch018

Raghav, A., Singh, B., Jermsittiparsert, K., Raghav, R., & Yadav, U. (2024). Artificial Intelligence in Environmental and Climate Changes: A Sustainable Future. In *Maintaining a Sustainable World in the Nexus of Environmental Science and AI* (pp. 485-506). IGI Global. DOI: 10.4018/979-8-3693-6336-2.ch019

Saini, K., Singh, A., Ahuja, A., Arora, N., & Saini, R. (2025). Research advancements in quantum computing digital twins. In *Digital Twins for Smart Cities and Villages* (pp. 37–53). Elsevier. DOI: 10.1016/B978-0-443-28884-5.00002-6

Sathianarayanan, M., Kirdemir, U., & Gianoli, A. (2025). Urban Intelligence: Navigating Climate Challenges With AI. In *Nexus of AI, Climatology, and Urbanism for Smart Cities* (pp. 59-90). IGI Global Scientific Publishing.

Satpathy, I., Nayak, A., & Jain, V. (2025). The Green City: Sustainable and Smart Urban Living Through Artificial Intelligence. In *Utilizing Technology to Manage Territories* (pp. 273-304). IGI Global.

Sepehri, B., Almulhim, A. I., Adibhesami, M. A., Makaremi, S., & Ejazi, F. Artificial Intelligence Role in Promoting Saudi Arabia's Smart Cities: Addressing SDGs for Socio-Cultural Challenges.

Singh, B., Kaunert, C., Arora, M. K., & Lal, S. (2025). Autonomous Transportation and Smart Vehicles Enhancing Mobility Solutions: Fueling IoT Solutions Achieving SDG 11 (Sustainable Cities and Communities). In *Sustainable Smart Cities and the Future of Urban Development* (pp. 67-90). IGI Global Scientific Publishing.

Sowjanya, C. S., Manoranjini, J., Needhu, C., Kulkarni, V. V., Merlin, S., & Nimmala, S. (2025). A Research Study on Concepts and Applications of Artificial Intelligence: Governance in Smart Cities. In *Sustainable Development Using Private AI* (pp. 1–23). CRC Press.

Suleimany, M., Gonbad, M. R. S., Naghibizadeh, S., & Niri, S. D. (2025). Artificial intelligence as a tool for building more resilient cities in the climate change era: A systematic literature review. *Artificial Intelligence and Machine Learning Applications for Sustainable Development*, 60-81.

Suryawanshi, S. S., Bhosale, D. G., Kshirsagar, V. G., Patil, D. P., Kapoor, M., Singh, S., & Patil, M. A. (2025). 12 Techno-Cities. *Advanced Manufacturing Processes: Integration of Sustainable Approaches*, 1, 189–208. DOI: 10.1201/9781003476238-12

Vashishth, T. K., Sharma, V., Sharma, M. K., & Sharma, R. (2025). Healthcare and Smart Cities Applications of Secure 6G Infrastructure. In *Building Tomorrow's Smart Cities With 6G Infrastructure Technology* (pp. 399-432). IGI Global Scientific Publishing.

Xu, G., & Guo, T. (2025). Advances in AI-powered civil engineering throughout the entire lifecycle. *Advances in Structural Engineering*. DOI: 10.1177/13694332241307721

Chapter 2
Harnessing Marketing Intelligence and AI to Understand Consumer Behavior in the Education Sector in Smart Cities

Amit Kumar Pandey
https://orcid.org/0000-0001-5767-6333
Institute of Management Studies, Lal Quan, India

Vijit Chaturvedi
Amity Business School, Amity University, Noida, India

Jaya Yadav
Amity University, Noida, India

D. S. Yadav
RBS Technical Campus, India

ABSTRACT

New-age marketing intelligence and AI are shaping the way the education sector harnesses consumer understanding in smart cities where digital infrastructure and technology integration is at its boom. AI-driven systems like predictive analytics and machine learning algorithms allow institutions to analyze large datasets, uncover correlations, and predict trends. For example, this data could help identify essential consumer segments such as students, parents, and educators, allowing them to target them with tailored marketing strategies. These kinds of smart city

DOI: 10.4018/979-8-3693-9410-6.ch002

using IoT devices and real-time data deliver adaptive education systems according to various urban demands. AI can empower the marketing decision-making process with greater accuracy and relevance in campaigns. Challenges like ethical data use, privacy concerns, and equitable access will have to be addressed. Leveraging AI-driven marketing/sales intelligence will include more audiences, drive creativity and innovation into education generation that will complement sustainability and developments of smart cities.

1. INTRODUCTION

Smart city innovations in education emerged as a key pillar that fosters quality of life, innovation, and economic growth. As urban environments have become more complex, understanding consumers in this sector has emerged as a strategic necessity. AI offers education providers the potential to critically assess consumer intelligence to offer smarter, targeted solutions. They are crucial to building consumer-based education systems that meet the diverse needs of urban populations in smart cities, where social infrastructure permeates with digital infrastructure and connectivity generates an ocean of information, with its flow the greater the number of agents involved. However, by leveraging marketing intelligence and AI, education sector stakeholders can create customized learning paths, manage resources more effectively, and improve overall engagement (Yadav & Shrawankar, 2025).

AI can improve marketing initiatives through hyper-personalization. AI can help institutions send personalized messages and content for different consumer segments, thereby increasing engagement and conversion rates. However, this is not limited to just the advertising space, let's see how we can improve our marketing with the help of artificial intelligence: for instance, if a language school is in the smart city, their artificial intelligence will identify all possible students interested in particular languages and the software then caters ads and courses directly to that specific audience (Salhab et al., 2025).

1.1 Understanding Consumer Behavior Through Marketing Intelligence

Marketing intelligence is the collection and analysis of data to know more about market trends, consumer behavior, and competitive landscapes. In the context of education, marketing intelligence refers to the collection of insights that help build a better educational offering based on the students, parents, educators, and institutions. In smart cities especially, as there are diverse demographics and top notch technology, consumer needs are always evolving. By harnessing marketing intelligence, educational institutions can better understand priority factors in students' decisions, which can range from affordability and accessibility to quality of education and technological integration (Hung et al., 2025). Parents might want economic affordability and location, students might look for experimental learning and global resources and the parents might want to ensure seniors get good jobs.

A marketing intelligence allows education providers to customize their services as per these varying needs. For instance, you can gain critical insights into consumer preferences throughout analyzing trends in enrollment, feedback, and dropout rates. The data obtained through these insights not only guide better crafting of marketing campaigns but also enhance the education delivery process. Marketing intelligence equips institutions with insights into emerging trends, such as the increased prevalence of online and hybrid learning models or skill-based certifications, enabling them to stay competitive in an ever-evolving market (Aishwarya, 2025).

1.2 AI-Driven Insights in Consumer Behavior Analysis

It further enhances marketing intelligence with real-time data analysis and predictive modeling. AI-based tools are capable of analyzing large volume of datasets from various streams, including social media, online platforms, IoT devices, and institutional archives. These tools detect patterns and trends that would be impossible for humans to identify manually. Machine learning algorithms, for example, can analyze data from smart city sensors and education platforms to predict patterns of enrollment, identify popular courses, and study the effectiveness of marketing campaigns. AI technology allows for a high level of personalization, critical to engaging consumers of urban education. From customized marketing messages to course suggestions and learning experiences, AI-enabled personalized recommendations enable institutions to provide tailored outreach. For instance, an intelligent AI-based platform could suggest courses based on students' learning history, interests, and career goals, resulting in a more personalized and relevant experience. Also, institutions can run sentiment analysis using AI, measuring consumer satisfaction and identifying possible areas of modification (Diaz-Llairo, 2025). By applying this knowledge, businesses can hone marketing techniques and advance service provision, resulting in increased retention and satisfaction rates.

1.3 Smart City Education for Supplementary Consumer Groups

Smart cities education sector has various consumers starting from the students to parents and educators and institutions. These different demographic generations have distinct preferences and expectations that need to be matched with specific approaches. As modern urban living demands continue to evolve, students are looking to enroll in more flexible learning models online courses, hybrid programs, micro-credentials and the like. Parents are more concerned about quality, safety, and price and generally make their decisions based on word-of-mouth, online reviews, and school rankings. Educators also are emphasizing professional development opportunities and access to state-of-the-art teaching tools. Institutions compete by statistics, offering the best and most competitive programs, by technology and by building their overall identity as an institutional brand. Understanding consumer behavior for these groups requires a multi-variate approach (Arantes & Costa, 2025). For example, institutions can employ AI to segment audiences and create personalized marketing efforts. For example, a smart city university may run targeted email campaigns to incoming students with information on course offerings, scholarships and career opportunities, while also tackling perspectives of parents, on whether or not the university is affordable and safe, through in-depth brochures and interactive webinars.

1.4 Obstacles to Implementing AI and Marketing Intelligence

Challenges in Adoption of AI and Marketing Intelligence in the Education Sector Although AI and marketing intelligence have transformative potential, there are multiple challenges in their implementation in the education sector. The most worrying is data privacy and ethical use of AI. In addition, the collection and analysis of consumer data must be done under rigorous safeguards to meet the requirements of data protection laws. Abuse of data can result in a breach of consumer trust and damage to the efficacy of marketing activities. The other major issue is the digital divide, which can create a hindrance to the equitable access to education in smart cities. Although AI and marketing intelligence can optimize delivery of information, they can also marginalize communities that do not have access to technology or digital literacy. The issue must be approached with a collaborative mindset that creates both systems that are inclusive and reflect socio-economic diversity (Al-Rimawi & Nadler, 2025).

These methods are costly infrastructure and resource-wise and that presents challenges to the adoption of AI within marketing intelligence. Research shows that smaller institutions may not have the financial and technical means to implement advanced AI solutions, resulting in a divide in the quality of education offerings. Moreover, the reluctance of established institutions and stakeholders to embrace change may hinder the incorporation of innovative technologies, necessitating efficient change management approaches (Merlo, 2025).

2. SMART CITIES, EDUCATION, AND MARKETING INTELLIGENCE: EDUCATION AS CRITICAL COMPONENT OF SMART CITY ECOSYSTEMS

In recent years, Smart cities have been one of the new-age solutions that aim to transform urban environments through the implementation of cutting-edge technologies, data, and interconnected systems. One of the many sectors that make up a smart city ecosystem, education is a major driver of innovation, society, and human capital development — and is subsequently a critical part of the large-scale overlap of industries that allows smart cities to thrive. With the world population in decline, developments in intelligent, adaptative and inclusive services for education have become unavoidable. In response to this demand, the fields of marketing intelligence and artificial intelligence (AI) are radically transforming the methods employing the ways education is provided, promoted, and consumed so that the education sector can adapt flawlessly to the dynamic needs of city populations.

That only changes if we address education which is fundamental in smart-city ecosystems where skilled labor is being developed, innovation nurtured, and civic engagement promoted. Education in a smart city goes beyond traditional boundaries and integrates concepts of technology, sustainability, and inclusivity into its framework. It underlies the preparation of citizen-makers to flourish in complex, tech-enabled urban places. Smart education ensures that the smart city acts as a center of lifelong learning, matching people with jobs and opportunity through the use of smart tools and systems. The future of smart cities, with interconnected systems, adaptable infrastructure, and responsive solutions, has huge implications for education systems (Ogunseye et al., 2025). Well before they become adults, schools are using technology to provide the kind of personalized, flexible and competency-based learning experience that they need from early childhood education through higher education and vocational training. Digital classrooms, online learning platforms, and hybrid education models are just a few examples of how smart cities are transforming the educational landscape. These developments bring not only access to quality education but also provide learning opportunities to different classes including the underserved and marginalized communities.

2.1 Marketing Intelligence in Education: Industry Marketing Intelligence

Marketing intelligence is the process of collecting and analyzing data in a structured manner to gain insights into consumer behavior, market trends, and competitive dynamics. Marketing intelligence in education is a powerful tool for gaining a deeper understanding of students, parents, and educators. This data can help educational providers build targeted strategies for their consumer base to maximize engagement, enrollment, and retention. In smart cities with interwoven data flows, marketing intelligence is a game changer for education providers. By studying data from a variety of sources, including enrollment numbers, student evaluations, digital engagement and social media trends, schools can detect patterns and adjust course accordingly (Alexopoulos et al., 2025). For example, across one university in a smart city, marketing intelligence might identify a growing interest in such courses related to AI, climate change or digital entrepreneurship. With this knowledge in hand, the organization can create and market programs that meet those interests and remain competitive in the education sector.

Marketing intelligence is also key to improving the student experience. Educational institutions that know their target audience allow them to develop personalized learning journeys, provide valuable resources and work on a marketing strategy that speaks to future leads. An example of this could be that data-driven insights indicate that pupils in a specific geographical location favour hybrid learning models which blend online flexibility with face to face depositions of learning. Institutions can leverage these insights to adjust course offerings and promotional strategies to match the demands and expectations of their target population. AI takes marketing intelligence to the next level through real-time data analytics, predictive modeling, and hyper-personalization (Tyagi & Anand, 2025). The utilization of AI-powered tools like natural language processing, sentiment analysis, and machine learning algorithms enable educational institutions to better understand consumers patterns and preferences. They are capable of analyzing these vast volumes of data from various sources like online learning platforms, social media, IoT devices, and institutional records, yielding actionable insights that inform decision-making.

The most valuable offerings of AI to educational marketing is personalization. The importance of personalized learning and marketing can be even more in smart cities, where there will be variety of populations with unique preferences, goals, and challenges. With its power of personalization, AI provides institutions with the ability to offer students tailored recommendations and course suggestions, as well as marketing messages, which helps to create a more relevant experience. An AI-powered platform, for instance, may suggest certain courses or resources to a student based on their academic records, career backings, and learning style. In a similar way, AI can scrutinize social media activity to locate potential students and reach out to them with tailored ads or content (Das et al., 2025).

The benefit AI has brought is predictive analytics, allowing for the ability to predict future trends and behaviors in the education field. Through analyzing historical data, AI can help predict enrollment trends for specific courses, identify emerging areas of student interest, and provide insight into how effective marketing campaigns have been. These predictions enable educational institutions to re-evaluate their strategies and keep pace in a hyper competitive marketplace. An AI model might, for example, predict a growing interest in courses related to renewable energy or cybersecurity, leading institutions to allocate resources in these areas and market such offerings to prospective students.

2.2 Consumers in Smart Cities and Education

Education sector in smart cities is a broad one as it serves a diverse range of consumers from students, parents, educators, and industry partners. Together, they have different demands, interactions, and expectations that need to be catered for accordingly in their engagement and service delivery. The students living in smart cities are looking forward to getting access to quality education that provides a flexible approach of learning through technological means while keeping industries trends in mind. And desire for opportunities to enhance their skills, grow their careers, and develop personally (Megha & Almeida, 2025).

The biggest concern parents have is affordability, in addition to the safety of the child and the general reputation of the educational institution. Teachers seek opportunities for professional development, access to innovative teaching tools and a supportive work environment. They value partnerships that not only produce a skilled workforce and enable innovation but also drive societal development. Institutions can use marketing intelligence and AI to effectively address these diverse consumer needs. For instance, they can utilize AI to segment their target audience and send personalized messages to distinct consumer groups. A smart city school, for example, could use data analytics to research parents' preferences in its vicinity, and tailor its marketing campaigns to emphasize affordability, safety protocols, and academic excellence. In tandem, the school might leverage AI-driven platforms to provide personalized learning recommendations for students, ensuring that their educational path reflects their ambitions (Pandey et al., 2025).

Marketing intelligence and AI in education can help address challenges.This is a significant concern in all fields, especially in smart cities, as these cities collect and share data across various systems that are interlinked. They should not yield to these over-incentives as institutions need to ensure that they comply with data protection laws and engage in ethical data collection and analysis. Another challenge depicted is the digital divide, where not all urbanites have the same access to technology and digital literacy. This has to be addressed by a commitment to diversify and address the differences between socio economic groups.

However, the potential for education in smart cities is huge and can help break down these challenges. Innovative tools like blockchain along with AR and VR can also help improve the education ecosystem. Blockchain to enhance transparency and security in educational records, and AR and VR to create immersive and interactive learning experiences. The combination of these technologies with marketing intelligence and AI can create education systems that are more engaging, more efficient and more consumer centric. Education plays a key role in smart city ecosystems, emerging as a driver of innovation, social progress, and human development. Marketing intelligence and AI used to understand and shape consumer behavior will allow educational institutions to adapt their content to the evolving world making it relevant, inclusive, and impactful. Though there do remain some obstacles needing to be worked on in this sector, such as data privacy and the digital divide, the potential for growth, change and transformation remains unmatched (Rosário & Cruz, 2025). In the future of smart cities, education will always be a priority for building robust, eco-friendly, and smart urban centers.

3. CONSUMER BEHAVIOR IN THE EDUCATION SECTOR OF SMART CITIES

Smart city education sector depicts a distinct overlap of high-tech, data-centric frameworks, and metropolitan life. Diverse factors from demographic diversity to technological advancement, and the dynamic of smart city ecosystems influence consumer behavior in this sector. The configuration of these factors determines the preferences, decision-making processes, and behavioral trends of all stakeholders in education — be it students, parents, educators, institutions, corporate training providers, etc. Only by understanding what these groups need, the expectations they have, how their behavior is changing, and what factors influence their decisions and trends in how they interact with the education sector can fair and sensible education systems be designed for smart cities (Hansen et al., 2025).

Smart cities have a wide range of consumer groups, each with a significant role, objectives, and expectations and the education sector needs to meet them. Students, parents, educators, institutions, and corporate training providers are among the key stakeholders, each of whom plays a vital role. Primary consumers in the education sector are students. Smart city students are also more tech-savvy, digitally literate and career-oriented than suburban or rural students. They thrive in environments where they can embrace the new and experience change. They want education systems that equip them for competitive job markets, especially in technology, artificial intelligence, sustainability, and entrepreneurship. In urban areas, too, students are also more likely to favor online and hybrid education models that are more convenient and adaptable to their schedules.

Factors like quality of education, safety, affordability, and technology integration in learning are priorities for the parents of smart cities. They consider the reputation of educational institutions, proximity to their campuses and the match between academic products and career prospects for their children when they make their decisions. Most urban parents are well-informed using digital tools to evaluate schools, keep tabs on their children's progress and directly connect with educators. In smart cities, educators play a dual role in the education sector as both consumers and providers. They want stufy abroad opportunities, use of modern Teaching technologies and enabling teaching environment with Innovative culture. Educators embody lifelong learners seeking to stay informed of new teaching practices and technological advances. As institutions in smart cities, they compete in an environment where students are the main clients, and therefore, they must attract and retain students (Jain, 2025). Their actions are influenced by the necessity to create, respond to market conditions, and provide a student and faculty experience of excellent quality. From technology integration and marketing intelligence, to data-driven decision-making and consumer behavior institutions are investing heaps of time and resources to truly "get to know it".

3.1 Corporate Training Providers

Corporate education trainers are a key segment of education sector trainers that cater to skill-based certifications, professional development courses, and industry-specific training. Smart cities hire these suppliers to target a range of employees looking for training and promotions. Back to their behavior, Showing a particular emphasis on shaping programs that fit into the employment needs of the urban industries like Info sec, green technology, and entrepreneurship (Sharma et al., 2025).

The smart city education consumer is nothing like the suburban or rural education consumer. Urban consumers tend to place emphasis on technology integration, digital infrastructure, and access to global markets. They place a higher value on innovation, convenience, and instant feedback, expecting state-of-the-art solutions and adaptable learning model. Their focus on rapid and responsive educational experiences drives the need for the hope of innovation in training and education, without a solid commitment to adherence to practices. Far more suburban and rural consumers may focus on affordability, locally based learning, and traditional educational framework. Such contrasting preferences illustrate the importance of tailored strategies that accommodate the distinct needs and limitations of these demographic groupings (Sumra et al., 2025).

Several of the key factors in consumer decision-making for the education sector of smart cities stem from consumer behaviour, with academic and institutional needs coming to the front across the terrestrial sector of the town across infrastructure within urban areas. In smart cities, accessibility is an important consideration for education consumers. This is both physical accessibility to campuses and digital accessibility to online resources. Physical access is improved through urban mobility solutions like public transportation and interlinked infrastructure, and educational content reaches the poorest through high-speed internet and digital platforms. As more consumers enter smart cities, one primary consideration is the integration of technology in education. Parents and students seek educational institutes that utilize advanced tools like virtual classrooms, Ai powered learning platforms and gamified content. Institutions that have access to the latest technologies allow educators to improve their teaching methods. Similarly, technology integration is also important for institutions wanting to remain relevant in a tech-centric urban world (Sowmya, 2025).

Education is still a top-of-mind issue for all three consumer groups. This span the relevance of curricula, the expertise of educators, and student achievement. Consumers expecting quality in smart cities expect it to be innovative, have global connectivity, and be able to adapt to rapidly evolving sectors. Urban consumers are particularly attracted to institutions offering experiential learning opportunities, industry partnerships, and global accreditation. Consumer behavior in the education sector is greatly affected by concern about affordability. However, as intelligent cities prompt higher education costs, the general urban consumers are looking for value for money even if they have a possible higher disposable income. Frequently, scholarships, financial aid, and suitable payment plans are vital in decision-making for students and parents.

Unique to smart city desire and in addition to general decision making influencers, consumer behavior is also affected by smart city consumer-related influencers. This includes urban mobility, digital infrastructure and the tech-savviness of the urban population. These allow for easier access to education due to interconnected systems in the area like transport, smart city applications and even IoT based services available. At the same time, the rise of tech-savvy customers demands Creative, data-driven, and personalized learning solutions.

4. AI-POWERED MARKETING INTELLIGENCE FOR SMART CITIES: DATA COLLECTION AND ANALYSIS IN SMART CITIES

They employ interlinked technologies and analytics-based frameworks to enhance urban living and evolve sustainable development. However, the most transformative tools used in smart cities are most likely Artificial Intelligence (AI) and marketing intelligence, which have the potential to collect and analyze huge amounts of data and transform that data into innovation and decision-making processes. It also utilizes advanced data gathering methods combined with deeper analytical techniques to create actionable insights for the real-time needs of societies, governments, institutions, and people across the world. Such measures not only improve operational efficiency but also offer more tailored experiences to citizens, promoting inclusivity and sustainability (Khazaei, 2025).

Smart cities marketing intelligence refers to the collection and analysis of data to gain insight into consumer behavior, preferences and emerging trends. It also enables stakeholders to make better-informed decisions and develop tailored strategies, which appropriately consider specific features of urban settings. Marketing intelligence in a smart city can go beyond traditional ones by applying them in areas like urban planning, healthcare, education, transportation, and environmental sustainability. AI-supported marketing intelligence systems are capable of analysing complex datasets originating from internal and external sources to discover patterns and predict future behavior, allowing stakeholders to take a proactive approach when facing problems or attempts to seize alternatives (Dey et al., 2025).

Marketing intelligence with AI is revolutionizing how data is collected and analyzed in smart cities. Utilizing cutting-edge algorithms and advanced machine learning models, big data can be analyzed in real-time revealing insights that are not just insightful but non-conceivable with traditional methods. As an instance, AI-enhanced marketing intelligence can process data from sensors, social media channels, mobile applications, and e-commerce websites to identify changing consumer needs in cities, forecast market requirements, and improve the resource allocation.

4.1 Smart City Data: IoT and Sensor Networks

The foundation of marketing intelligence is thorough data collection, and smart cities offer a rich environment for one to collect heterogeneous and rich data. Smart cities have a complex architecture where different pieces of data from many sources are generated such as IoT devices, sensors, mobile applications, social media, and public records. This applies to all aspects of data collection, which is made more efficient, automated, live, trustworthy and concrete by AI. Smart cities rely on IoT devices and sensor networks to collect data on urban systems, the environment, productivity, citizens, mobility, and government to maintain the intelligent city ecosystem. Traffic flow, pollution, energy use, public safety, etc., are just some of the data collected by these devices, offering us qualitative insights into the dynamics of urban populations. Sensors embedded in smart buildings, for example, can monitor energy consumption and human activity, paving the way for customized marketing campaigns that promote energy-saving devices or eco-friendly behaviors (Praveen & Kumari, 2025). Traffic sensors, for example, can send data to transport companies who can now use that data to optimise individual routes, improving customer experiences.

Smart cities have an abundance of mobile apps and digital platforms, which continuously generate a wealth of data about users' preferences, habits, and interactions. Whether for e-learning or ride-hailing, these platforms store data on user behavior, location data, engagement metrics, among others that can be analyzed to create personalized marketing strategies. This data is processed by AI algorithms to discover trends, segment audiences, and suggests customized solutions. For example, an intelligent education app in a smart city can analyze how users engage with different courses and recommend personalized learning paths based on their unique data. A wealth of consumer data can be found on social media platforms, providing insight into consumer preferences, opinions, and social trends. AI is used in smart cities to parse interactions in social media for emergent demand, patterns of sentiment and the impact of influencers (Farooq et al., 2025). It provides valuable insight into how brands and institutions can speak to urban consumers who are engaged with digital. Use Case 1: Social Media Listening for Traditional Education Institutions

4.2 Public and Open Data Sources

In smart cities, public records and open data initiatives add extra layers of information to marketing intelligence. Demographics, employment, education, education levels, populations and public service. These data can be linked to other data sets to give an overall picture of urban populations, detailed monitoring, and tracking of urban populations over time. AI can analyse this data to identify correlations and make predictions about future developments so that stakeholders can craft strategies that meet community needs. Data collection offers raw material, however the actual value comes with evaluation which presents the data into actionable insights. This data should also be clean with no outliers so that the AI can come up with the marketing intelligence you need to generate the needed insights (Xiaoxin et al., 2025). Key elements of this process include data cleaning, data integration, data segmentation, and predictive modeling.

4.3 Data Cleaning and Integration: Segmentation and Personalization

Data extracted from multiple sources is commonly inconsistent, redundant or erroneous. With the help of AI-based tools, your data cleaning will be automated, making it more accurate and reliable. Once the data is cleaned, data from various sources is merged into a single view of the consumer behavior. As an example, an AI system can use data from IoT devices, social media platforms and public records to create a comprehensive profile of city consumers Thus, segmentation is one of the most important steps in the data analysis for marketers, enabling them to segregate urban populations into different groups depending on their demographics, preferences, or behaviors. It helps segmentation by analysing large and diverse datasets to detect subtle patterns that traditional means may miss. This allows the development of extremely tailored marketing campaigns (Singh et al., 2025). As an example, an AI system could break urban learners into segments based on their interest in a STEM field groups and provide specialized course or certification to each group.

Predictive modeling is one of the most potent capabilities of AI for marketing intelligence—it relies on historical data to predict future behaviors and trends. Within smart cities, predictive models can provide advance notice for shifts in consumer preferences so that stakeholders can catch the wave rather than chase it. For instance, an AI system may identify a trend towards hybrid education delivery models, leading to decisions to invest more in digital infrastructure and online learning courses. NLP tools process textual data, including social media posts, customer reviews, and surveys, to gauge consumer sentiment and preference. In smart cities, it helps obtaining valuable insights about public opinions and brand perception. NLP tools, for instance, can be used to analyze social media conversations regarding public transportation systems, revealing areas for improvement and aiding in marketing strategies (Shafik, 2025).

5. CHALLENGES IN IMPLEMENTING AI AND MARKETING INTELLIGENCE

Artificial Intelligence (AI) and marketing intelligence has brought a tremendous potential for evolving urban learning ecosystems to smart cities like never before. But, these changes come with challenges, though. However, there are many ethical, infrastructural, and behavioral challenges to integrating these advanced technologies into everyday processes. These challenges need to be comprehended and alleviated in order to harness AI-powered marketing intelligence within smart cities properly.

5.1 Ethical and Privacy Concerns

Data collection and consumer profiling, which goes hand in hand with implementing AI and marketing intelligence in the education sector of smart cities, is one of the hottest challenges in terms of its ethical and privacy-related risks. Some smart cities use the power of huge data generated from IoT devices, online platforms, and digital infrastructures. As such, this data is critical for understanding consumer behavior and preferences, however, it poses serious concerns regarding privacy violations, data misuse and unauthorized profiling (Grabocka & Ndoka, 2025).

5.2 Potential Pitfalls of Collecting Data and Building a Consumer Profile

EdTech systems that are powered by AI marketing intelligence systems often gather sensitive information relating to students, parents and educators for e.g., academic performance, learning preferences and behavioral patterns. Though useful for customization, such data collection may have side effects, including data breaches, identity theft, or discrimination. Similarly, when AI algorithms are not constructed with fairness and transparency in mind, it may unintentionally exacerbate biases within the system and create disparities in the access that various groups have to learning materials or opportunities.

The education sector in smart cities must balance these opportunities with regulatory frameworks that govern how and where data is collected, housed and used. Governments and institutions must also comply with strict data protection laws, which require transparency and accountability, like the General Data Protection Regulation (GDPR) in Europe that governs the use of personal data. Finding the right balance between the need for actionable insights and the imperative to safeguard individual privacy is a crime left to the devil. The varying rules and regulations across various regions can exacerbate the challenge for global education platforms in the operation of smart cities (Rana & Bhambri, 2025).

The major hindrances to having AI and marketing intelligence in smart cities are infrastructural and resource-related limitations associated with deploying these technologies. All smart cities are not created equal. Indeed, while some urban areas have the means to modernize their systems with state-of-the-art digital infrastructure and efficient internet of things networks, in other areas particularly within developing regions outdated technology, unreliable internet connections and a lack of digital awareness among stakeholders remain hurdles. Such inequalities contribute to a digital divide that impedes the fair implementation of AI and marketing intelligence within the educational sphere. Despite the promise of AI to increase educational efficiency, there are limitations that may arise in smart cities, especially in rural or underdeveloped areas with little infrastructure to support AI-enabled learning platforms, which most likely leads to marginalizing certain populations (Qadri et al., 2025).

5.3 High Costs and Technical Know-How as Barriers for Small Institutions

Smaller educational institutions and marketing startups may find these systems are cost prohibitive, which are some of the high costs often associated with implementing AI, marketing intelligence systems. This cost is not just limited to purchasing AI tools or IoT devices, it also covers digital infrastructure maintenance and hiring skilled professionals to run these systems. And the deep technical knowhow needed to deploy and operate AI solutions typically resides in a small number of personnel, compounding the gap in resources. Consequently, it becomes harder for smaller institutions to match up against larger groups, leading to unequal access to sophisticated assets in education.

6. FOCUSING ON CONSUMER-CENTRIC EDUCATION MODELS: BUILDING MARKETING STRATEGIES BASED ON REAL-TIME DATA INSIGHTS

Consumer-oriented education models are transforming the education space across smart cities. These models emphasize learner and stakeholder (e.g. parents, educators, and institutions) needs, interests, and behaviours. Advanced technologies like Artificial Intelligence (AI), the Internet of Things (IoT), and big data analytics make it possible to have real-time data insights which impact the marketing strategies developed to make consumers the focus of the marketing efforts. This will enable smart city education providers to develop individualized, efficient, and inventive methods that meet the varied needs of city dwellers by using these insights. The institution-led programs are being replaced with learner-driven approaches as the focus of consumer-centric education models. To better align with the needs of individual learners, these models focus on personalisation, accessibility, and adaptability in education (Malhotra & Kharub, 2025). Consumer-centric models have become increasingly relevant in smart cities, which rely heavily on digital infrastructure and technological innovation as core pieces of urban life.

The models are designed to offer value to learners by catering to their specific needs learning at their own pace, ability-based certifications, and customized content delivery. To illustrate, in smart cities, whichever working professional can prefer courses that are short and specific to a work area, and delivered over mobile platforms, while school students may benefit from gamified education platforms that make learning fun and interactive. As real-time data is generated by learners and other stakeholders, understanding these diverse needs can only result from deep analysis of this large data set (Cities, 2025).

Effective marketing strategies for consumer-led education models depend on real-time data insights. Education providers gain insights into their target audience's preferences, behaviors, and expectations, which allows them to create targeted and personalized campaigns. In smart cities IoT devices, digital learning platforms and mobile applications create massive amounts of data. For example, learning management systems (LMS) record metrics like course completion rates, time spent on particular modules, and performance of assessments. Social media networks and online forums are another source of insight into consumer attitudes and preferences, notes Oracle. By scrutinizing this data on a real-time basis, education providers can spot trends, predict requirements, and tweak their marketing plans accordingly (Gupta & Bhatnagar, 2025).

Powered by real-time data insights, highly personalized marketing campaigns can be developed that speak directly to the individual learner. AI tools, for instance, can suggest courses or programs based on a student's performance history or recommend additional resources that align with their overall career goals. Likewise, using parents' preferences data, schools or educational tools can then be showcased to the respective parents explaining the benefits of it. This allows education providers to create dynamically tailored content that is much greater than simply personalized learning it also drives a deeper relationship between education provider and audience. AI powered predictive analytics also make it possible for education providers to anticipate future trends and behaviours based on historical data. For example, an analysis of enrollment patterns could show growing interest in emerging fields like artificial intelligence, data science, or sustainable development. Smart city trends are expected to impact higher education institutions in urban areas significantly (Mohapatra, 2025).

The marketing strategies work well when the market needs are divided into consumer groups through real-time data insights. For smart cities, these consumer groups can be broadly categorized as students, parents, educators, and corporate training providers (Lohani et al., 2025). Different groups have different needs, and they expect the world to customize use of these technologies to them. Students, for instance, may pose an emphasis on affordability and flexibility, while corporate training providers may ask about alignment with policies in the industry (Singh et al., 2025). In smart cities, hardware is ubiquitous and hence the internet may be the best medium to market your educational offerings. Specific audiences can be reached through social media, email campaigns, as well as search engine optimization (SEO) strategies (Mouazen et al., 2025). Leveraging the insights derived from real-time data, these channels can be optimised by pinpointing the best content, the best timing, and the best formats to engage with (Sowjanya et al., 2025).

The focus of a consumer-centric approach is to provide seamless and engaging user experiences. Features that increase accessibility, like mobile-friendly platforms, multilingual content, and adaptive learning tools, should be emphasized (Arora et al., 2024). Real-time data is utilized to track user feedback for continuous improvement of the quality of educational services. The Importance of Consumer trust In Education marketing strategies appointment as it can use real-time data insights to showcase the value of educational offerings through testimonials, case studies, and performance metrics. QAI education software can help drive the academic progress of millions of learners, but it requires the confidence of consumers in its fairness, safety, and ethics (Raghav et al., 2024). Transparent policies and ethical practices further strengthen trust, ensuring that consumers feel confident interacting with AI education providers. Therefore, the marketing strategies must include feedback loops through which consumers can voice their opinions and feedback. Additionally, real-time analysis of feedback can help gain insights that can be practically applied for enhancing marketing efforts and strengthening educational programs (Gupta et al., 2025). If learners are unhappy about a program, for instance, providers can respond in real time, fixing the points of failure and announcing changes to an audience so those complaints are unlikely to drown them out (Sreedharan et al., 2025).

7. CONCLUSION AND FUTURE DIRECTIONS

The intersection of AI and marketing intelligence is an untapped potential for growth for smart city education stakeholders. AI can assist institutions in developing adaptive learning systems that suit various ways of learning as well as different requirements. AI-based platforms, for example, can offer real-time feedback to students, allowing them to monitor their development and discover areas for improvement. These platforms may also benefit educators by automating administrative tasks, giving them more time to focus on teaching and mentoring. The other big opportunity, is the use of AI for Predictive Analytics. AI can use past data to predict future trends in education, such as skills that employers will be looking for in the job market. It can help institutions create programs that cater to these needs, thus making sure that students are better suited for their job roles.

The role of AI and marketing intelligence in the education sector is not going away anytime soon as the smart cities of the future come to life. Future avenues of research should include more in-depth examinations of creating ethical guidelines for the use of AI in education marketing, as regards data privacy and consumer rights protection. We need to work towards systems that are inclusive and close the digital divide there has to be more equitable access to education. Blockchain, augmented reality, and virtual reality are some new-age technologies that can drive education in smart cities. Blockchain offers transparency and security for educational records, while AR and VR enable immersive experiences. When these technologies are combined with AI and marketing intelligence, engagement with consumers and satisfaction with their needs increase. Leveraging additional marketing intelligence and AI in the sector of smart city education is an effective way to comprehend and respond to consumer behavior. As such, education providers can harness these technologies to provide personalized, efficient, and inclusive systems that cater to the diverse needs of urban populations. Nonetheless, it is essential to address issues like data privacy, digital divide, and resistance to change for fully harnessing the power of these innovations. The continuation of this collaborative process, guided by research and with the involvement of all relevant actors, will be key to creating more sustainable, efficient, and consumer oriented education systems in smart cities.

REFERENCES

Aishwarya, R. I. (2025). 28 Smart Cities for the Future. *Cybersecurity and Data Science Innovations for Sustainable Development of HEICC: Healthcare, Education, Industry, Cities, and Communities*, 406.

Al-Rimawi, T., & Nadler, M. (2025). Leveraging Smart City Technologies for Enhanced Real Estate Development: An Integrative Review. *Smart Cities*, 8(1), 10. DOI: 10.3390/smartcities8010010

Alexopoulos, C., Saxena, S., Matheus, R., & Rizun, N. (2025). Artificial Intelligence (AI)-powered public service delivery in smart cities: The case of the Gulf Cooperation Council (GCC) Countries. In *ICT, the business sector and smart cities* (pp. 85-98). Routledge.

Arantes, L., & Costa, S. (2025). Converging Theories: The Fusion of Artificial Intelligence and Sustainable Marketing for a Responsible Future. In *Leveraging AI for Effective Digital Relationship Marketing* (pp. 223-256). IGI Global.

Arora, M. K., Lal, S., Singh, B., & Olyaee, M. (2024). Through an Energy Lens: Examining SDGs From a Policy Perspective. In Riswandi, B., Singh, B., Kaunert, C., & Vig, K. (Eds.), *AI Applications for Clean Energy and Sustainability* (pp. 320–343). IGI Global. DOI: 10.4018/979-8-3693-6567-0.ch016

Cities, S. (2025). 7 AI-Optimized Urban Resource Management for Sustainable. *Financial Inclusion and Sustainable Development in Sub-Saharan Africa*, 96.

Das, D., Bapat, J., Katsenou, A., & Shrestha, S. (2025). Connectivity for AI enabled cities—A field survey based study of emerging economies. *arXiv preprint arXiv:2501.09479.*

Dey, D., Majumder, A., Agrawal, Y., Tewari, S., & Mohapatra, H. (2025). Smart Mobility Revolution: Harnessing IoT, Sensors, and Cloud Computing for Intelligent Automobiles in the Urban Landscape. In *Sustainable Smart Cities and the Future of Urban Development* (pp. 143-164). IGI Global Scientific Publishing.

Diaz-Llairo, M. A. (2025). Exploring a Smart City and the Epicenter of Megatrends in Innovation With Artificial Intelligence: The Case of Dubai. In *Data-Driven Governance Through AI, Digital Marketing, and the Privacy Interplay* (pp. 191-238). IGI Global Scientific Publishing.

Farooq, M., Ramzan, M., & Yen, Y. Y. (2025). Exploring the Impact of Artificial Intelligence on Brand Management.

Grabocka, E., & Ndoka, E. (2025). AI-driven innovation within the ICT sector. *Smart Cities and Regional Development (SCRD)Journal*, 9(1), 77–97.

Gupta, M., & Bhatnagar, R. (2025). Harnessing AI in Geospatial Technology for Environmental Monitoring and Management: Applications of AI for Geospatial Data Processing. In *Recent Trends in Geospatial AI* (pp. 73-102). IGI Global Scientific Publishing.

Gupta, M., Deswal, P., Chauhan, V., Kumar, P., Kaur, J., & Singh, P. (2025). Optimizing Sustainability: Harnessing AI for Personalized Electric Vehicle Energy Management. In *Optimized Energy Management Strategies for Electric Vehicles* (pp. 233-260). IGI Global Scientific Publishing.

Hansen, M. M., Koonsanit, K., & Kulmala, V. (2025). How can data contribute to Smart City innovation: A study from Thailand's Smart City initiatives. *Frontiers in Sustainable Cities*, 6, 1473123. DOI: 10.3389/frsc.2024.1473123

Hung, T. H., Pandey, A., Mittal, S., & Chandel, A. (2025). Harnessing IoT and geomarketing for agile marketing. In *Digital Transformation Initiatives for Agile Marketing* (pp. 167–186). IGI Global.

Jain, R. (2025). Integrating Artificial Intelligence for Smart Destination Management in the UAE: Opportunities, Challenges, and Impacts. *Advancing Smart Tourism Through Analytics*, 245-260.

Khazaei, P. (n.d.). Evaluating the Role of Artificial Intelligence in Crafting Emotional Advertising and Enhancing Customer Experience in Diverse Industry Sectors.

Lohani, U., Gupta, D., Sivadas, S., Gond, P., Kumar, D., Attwal, K. P. S., & Misra, V. (2025). Harnessing the Internet of Behavior (IoB) for Food Safety and Quality: A Conceptual Approach. In *Mapping Human Data and Behavior With the Internet of Behavior (IoB)* (pp. 271-300). IGI Global Scientific Publishing.

Malhotra, G., & Kharub, M. (2025). Elevating logistics performance: Harnessing the power of artificial intelligence in e-commerce. *International Journal of Logistics Management*, 36(1), 290–321. DOI: 10.1108/IJLM-01-2024-0046

Megha, C. T., & Almeida, S. M. (2025). Blossoming Fields: Harnessing Artificial Intelligence for Smart Agritourism. In *Cases on AI-Driven Solutions to Environmental Challenges* (pp. 213-240). IGI Global Scientific Publishing.

Merlo, T. R. (2025). The Dynamic Nexus of Artificial Intelligence and the Co-Creation of Smart Cities Innovations: The Brazilian Ranking. In *Advancing Smart Tourism Through Analytics* (pp. 423-434). IGI Global Scientific Publishing.

Mohapatra, H. (2025). The Role of 6G in Empowering Smart Cities Enabling Ubiquitous Connectivity and Intelligent Infrastructure. In *Building Tomorrow's Smart Cities With 6G Infrastructure Technology* (pp. 237-264). IGI Global Scientific Publishing.

Mouazen, A. M., Hernández-Lara, A. B., Chahine, J., & Halawi, A. (2025). Triple bottom line sustainability and Innovation 5.0 management through the lens of Industry 5.0, Society 5.0 and Digitized Value Chain 5.0. *European Journal of Innovation Management*. Advance online publication. DOI: 10.1108/EJIM-04-2024-0339

Ogunseye, O. O., Ajayi, O. T., Fabusoro, A., Abba, A. O., & Adepoju, B. (2025). Leveraging Artificial Intelligence for Advancing Key Sectors of National Growth and Development. *Asian Journal of Current Research*, 10(1), 45–55. DOI: 10.56557/ajocr/2025/v10i19056

Pandey, A., Yadav, M., & Hung, T. H. (2025). Enhancing Customer Experience Through Geomarketing and IoT. In *Digital Transformation Initiatives for Agile Marketing* (pp. 187–218). IGI Global.

Praveen, N., & Kumari, N. (2025). Machine Learning Concepts in Education and Skill Development: Cloud Solutions for Smart Cities. In *AI and Green Technology Applications in Society* (pp. 219-248). IGI Global Scientific Publishing.

Qadri, U. A., Moustafa, A. M. A., & Abd Ghani, M. (2025). They misused me! Digital literacy's dual role in AI marketing manipulation and unethical young consumer behavior. *Young Consumers*. Advance online publication. DOI: 10.1108/YC-08-2024-2207

Raghav, A., Singh, B., & Jermsittiparsert, K. (2024). Role of Artificial Intelligence and Machine Learning in Clean and Blue Energy for Enhancing Efficiency and Sustainability. In Riswandi, B., Singh, B., Kaunert, C., & Vig, K. (Eds.), *AI Applications for Clean Energy and Sustainability* (pp. 366–386). IGI Global., DOI: 10.4018/979-8-3693-6567-0.ch018

Rana, R., & Bhambri, P. (2025). Harnessing AI and Machine Learning for Enhanced Geospatial Analysis. In *Recent Trends in Geospatial AI* (pp. 27–72). IGI Global Scientific Publishing.

Rosário, A. T., & Cruz, R. N. (2025). Data-Driven Decision-Making in Marketing.

Salhab, H., Zoubi, M., Khrais, L. T., Estaitia, H., Harb, L., Al Huniti, A., & Morshed, A. (2025). AI-Driven Sustainable Marketing in Gulf Cooperation Council Retail: Advancing SDGs Through Smart Channels. *Administrative Sciences*, 15(1), 20. DOI: 10.3390/admsci15010020

Shafik, W. (2025). Artificial Intelligence-Enabled Cybersecurity and Internet of Things Applications in Smart Cities. In *Building Tomorrow's Smart Cities With 6G Infrastructure Technology* (pp. 301–334). IGI Global Scientific Publishing.

Sharma, S., Thirumalaisamy, M., Balusamy, B., & Chilamkurti, N. (Eds.). (2025). *Artificial Intelligence and IoT for Cyber Security Solutions in Smart Cities*. CRC Press.

Singh, B., Kaunert, C., Lal, S., Arora, M. K., & Jermsittiparsert, K. (2025). Intelligent Mobility Assimilating IoT in Autonomous Vehicles: Foster Sustainable Cities and Communities. In *Designing Sustainable Internet of Things Solutions for Smart Industries* (pp. 279-300). IGI Global.

Singh, S., Madaan, G., Swapna, H. R., Singh, A., Pandey, B. K., George, A. S., & Pandey, D. (2025). Unleashing the Power of AI and Data Analysis: Transforming Insights into Action. In *Interdisciplinary Approaches to AI, Internet of Everything, and Machine Learning* (pp. 1-24). IGI Global Scientific Publishing.

Sowjanya, C. S., Manoranjini, J., Needhu, C., Kulkarni, V. V., Merlin, S., & Nimmala, S. (2025). A Research Study on Concepts and Applications of Artificial Intelligence: Governance in Smart Cities. In *Sustainable Development Using Private AI* (pp. 1–23). CRC Press.

Sowmya, D. S. (2025). Case Studies and Real-World Applications. In *Mapping Human Data and Behavior With the Internet of Behavior (IoB)* (pp. 525–554). IGI Global Scientific Publishing.

Sreedharan, S., Ramachandran, M., & Ramesh, D. (2025). Harnessing digital twins and industrial-IoT for cutting-edge mining automation: A methodological and technology assessment prototype. *Computers & Industrial Engineering*, 201, 110871. DOI: 10.1016/j.cie.2025.110871

Sumra, K. B., Siddique, H., Afzal, S., & Qazi, A. (2025). Sustainable smart cities: promotion of circular economy in urban GCC regions. *Journal of Science and Technology Policy Management*.

Tyagi, U., & Anand, N. (2025). Solar Energy and Smart Cities: A Holistic Approach for Sustainable Development. In *Renewable Energy Development: Technology, Material and Sustainability* (pp. 91–109). Springer Nature Singapore. DOI: 10.1007/978-981-97-9626-7_5

Xiaoxin, L., Chuah, S. H. W., & Lye, V. W. E. (2025). Harnessing the Power of Social Media Marketing to Transform the Food Industry in China. In *Technological Innovations in the Food Service Industry* (pp. 23–48). IGI Global Scientific Publishing.

Yadav, U., & Shrawankar, U. (2025). Artificial Intelligence Across Industries: A Comprehensive Review With a Focus on Education. *AI Applications and Strategies in Teacher Education*, 275-320.

Chapter 3
Integrating Mobility in Multimodal Transportation Systems: A Comprehensive Analysis and Their Impact on Urban Mobility

Manmeet Kaur Arora
https://orcid.org/0009-0002-5071-117X
Sharda University, India

Sahil Lal
https://orcid.org/0000-0001-9827-3717
Sharda University, India

Bhupinder Singh
https://orcid.org/0009-0006-4779-2553
Sharda University, India

Saquib Ahmed
https://orcid.org/0009-0008-1891-6910
Sharda University, India

Christian Kaunert
https://orcid.org/0000-0002-4493-2235
Dublin City University, Ireland & University of South Wales, UK

ABSTRACT

As cities face the escalating demands of urban mobility, sustainability, and accessibility, multimodal transportation systems are gaining recognition as a solution to

DOI: 10.4018/979-8-3693-9410-6.ch003

these complex challenges. It examines how different modes of transport (cars, public transport, bicycles, walking, and micromobility) can be integrated into seamless transport networks which improve the user experience as well as the efficiency of transport operations. The chapter outlines the challenges facing multimodal systems, including complex iteration requirements, coordination between the numerous different transportation providers, differing regulatory environments, and the need for users to adapt to integrated systems. It goes on to highlight best practices, keeping users at the centre of the agenda when introducing new integrated ticketing solutions and developing robust infrastructure to support such systems in place. It concludes with policy recommendations, stressing the need for regulatory alignment, sustainable funding frameworks, and incentives for sustainable practices.

INTRODUCTION

Multimodal transportation is a groundbreaking mode of transport, providing the accessibility and interconnectivity to help the passage of people and goods. This system recognizes that no individual mode can serve all transportation needs effectively, therefore striving for a mix of transportation modes, including walking, cycling, public transportation, and private vehicles. The idea behind Travelled is based on the belief that improving accessibility and transportation, is only possible by taking the big picture of the transport world through the connection between different transport methods. Multimodal transportation seeks to combine not only the strengths of individual modes of transport but also its supplementary benefits to fullfill the needs of mobility solutions in the urban and regional context effectively. With the rising urbanization, population density, and environmental challenges, multimodal transportation has gained significant importance. As urban centres grow and traffic congestion worsens, legacy car-first planning paradigms no longer are up to the task. Multimodal systems, which promote multifaceted transportation methods, Dass relieved traffic, lowered emissions, and increased the ambiance of living and disposition of the inhabitants. Additionally, multimodal transport facilitates economic development by expanding access to employment, services, and recreation (Toshniwal, 2024).

Mobility and Accessibility Multimodal transportation planning is inherently about mobility and accessibility. Mobility describes the potential of people to travel from origins to destinations with different modes of transportation, and accessibility measures how well agents can access desired activities and services distributed over geographic space (Sharma, 2024). Core features. The conceptual boundaries of multimodal transport are drawn along a few important principles: Context-sensitive approach looking at the larger transportation ecosystem Holistic coordination of land use and transportation planners Cooperative inter-governmental partnerships Full consideration of pedestrian, transit, bicycle and automobile travel (Kalawatia & Rajeraman, 2024).

Broadly, multimodal transportation simply means utilizing more than one transport method in a single journey or a logistics operation. This can include combinations of roadways, railways, waterways, and air transport. And unlike intermodal transportation—where different modes are involved under different contracts multimodal transport is generally led under one contract with a single service provider that is accountable for the full trip. This distinction is important as it highlights the integrated nature of multimodal systems which intend to smooth out tasks and use efficiency. Several fundamental concepts underlie the framework for multimodal transport:

Integration: A fully effective multimodal system needs the components of such to be integrated. This extended to the physical infrastructure, such as transit hubs and interchanges that enable seamless mode transfers.

Convenience: The primary aim of a seamless multimodal transport system is to provide greater convenience for all users, including those without private vehicles. This means anticipating all the different user types and ensuring access to all modes for people with disabilities.

Multimodal systems promote sustainability by encouraging green transportation choices, such as cycling, public transit, and walking. As such, these systems contribute to reduction of greenhouse gas emissions and air quality improvement by reducing the dependence on single occupancy vehicles. Data-Driven Decision Making, Multimodal systems should leverage data analysis to improve network efficiency, optimize travel routes, and better match supply with demand. When done correctly, this user-centric approach allows people to use different modes depending on the situation and their needs.

Sustainable: Transportation agencies need to explore multimodal transportation options and collaborate between solution providers as well as connect with the actual assets to ensure successful implementation across public infrastructure. Such collaborative work guarantees that a range of voices are involved in planning processes.

Fundamentals of Multimodal Systems

In order to bring a multimodal transportation system to fruition, a few important aspects need to be kept in mind:

Harmlessness: Most transport does not cause slowdown problems, other than train transport. This includes multimodal transit stations, bike lanes, sidewalks, and parking facilities designed for multiple modes. Suitable infrastructure should be in place and in the right locations, maximizing accessibility and connectivity.

Policy Context: Supportive policy development environment to scale multimodal initiatives. Policies should promote investment in diverse transport modes and tackle regulatory barriers that would otherwise prevent integration.

Factors that have an impact in upgrading multimodal systems are: Technology Real-time information systems inform users of schedules and delays of all modes. Integrated payment systems also make it easier to collect fare amongst different transport providers.

Multimodal Planning: Exploring transit alternatives requires an understanding of community needs and preferences, making public engagement activities a key component of place making. A key need of non-profits, foundations and public agencies is a public outreach that collects useful input, which informs decision-making and builds support for the programs of their knowledge economy initiatives.

Safety: Safety is the top concern when it comes to any transportation system. Safety must be baked into multimodal planning in a way that protects users regardless of mode, such as providing better signage at crossings or adding lighting in transit spaces.

The systems must have sustainable funding methods to maintain multimodal infrastructure and services over time. Financing collaborations could involve public-private partnerships or dedicated funding streams from local government.

Multimodal transportation is a transformation in the way we solve mobility problems in urban areas. When different types of transportation are combined into a seamless network that focuses on accessibility, sustainability, and user experience, communities can increase their overall quality of life while also solving common problems like congestion and environmental degradation. This approach is grounded in a conceptual framework that stresses the significance of cooperation between stakeholders but recognizes the necessity for supportive policies and strong infrastructure to unlock the complete benefits of multimodal frameworks. However, as we are now on the verge of new demographic trends and technological changes, cities must embrace multimodal transportation systems to build resilient urban environments that can meet the diverse needs of their residents today and tomorrow (Smith & Johnson, 2023).

Walking

Walking is the most basic form of transport and plays a key role in multimodal systems. It is sustainable by nature and demands less infrastructure than other modes. Walking fosters connectivity to community and health through movement. To make walking a feasible option for transportation, cities should be designed with pedestrian-friendly infrastructure like sidewalks, crosswalks, pedestrian signals, and safe street crossing areas. Connecting walking paths to public transit stations facilitates multimodal trips. Still, there are obstacles to encouraging walking as a primary transport mode. Pedestrians are often at the mercy of the vehicles that dominate urban spaces, with consideration for their experience falling by the way-side; thus it's common to see unsafe or relatively unwelcome walking conditions. By designing better infrastructure for pedestrians, we can improve walkability that not only helps pedestrians but are also essential for the effectiveness of multimodal transport systems.

Micromobility Options

Innovative solutions have been developed in the urban transportation landscapes such as micromobility options. This include e-bikes, electric scooters (e-scooters), skateboards, and other lightweight personal vehicles meant for short-distance travel Micromobility has provided users with flexible options that can also supplement traditional public transportation systems by enhancing last-mile connectivity. Image: Savioe Micromobility has seen incredible growth in recent years, made possible by advances in technology alongside shifting consumer preferences to more sustainable transport options. Often, these vehicles are made accessible to users through shared services where they can rent them on-demand using mobile applications. Micromobility also offers flexibility, making it attractive in crowded metropolitans where owning a car might prove difficult. Even with their advantages, micromobility comes with drawbacks such as safety regulations, infrastructure requirements (like designated lanes), and potential conflicts with pedestrians or other traditional vehicles. This can be part of you overall policy framework how to deal with these problems and also integrate these in the multi modal transport systems that make sure these are all dealt with in a safe way.

Integration Strategies

To combine various transportation modes into one, successful integration of these various systems is critical to creating a harmonious full multimodal system that serves every class of user. Some strategies for integration include:

Integrated Infrastructure Planning: Designing integrated infrastructure that interchanges multiple modes e.g., bike lanes next to bus stops or sidewalks leading to transit hubs — allows for smooth transfer between various transport systems.

Unified Payment Systems: The introduction of integrated fare systems that enable users to pay using a single payment method or card across varying modes

Real-Time Information: Information on schedules and availability in real time along all means of mobility contribute to user convenience and promote multimodal travel.

Community Education: Raising public awareness of the advantages of multimodality can help increase acceptance and promote changes in transportation habits.

Infrastructure Considerations

Infrastructure is a vital modality for the success of multimodal transportation systems. Port of entry planning must take into consideration the unique infrastructure requirements of each mode and work to ensure compatibility between them. Dedicated bike lanes must connect to the public transit facilities so that cyclists can transition safely to buses or the train. Some important factors to consider when planning a cloud infrastructure are:

Design Standards Developing design standards that set a minimum standard for the safe and accessible accommodation of different modes. This might involve curb cuts for wheelchair users, clearly marked crosswalks for pedestrians, or appropriate signage for cyclists.

Interchange Facilities: Interchange facilities are an integral part of multimodal systems where users transition between alternate modes. Such facilities should be designed for efficiency, with things like sheltered waiting areas, secure bike parking, and easy transfers to transit vehicles.

Maintenance: Regular maintenance is key to keeping infrastructure safe and usable over time. Deterioration that interferes with user accessibility or safety can usually be prevented by regular inspections and timely repairs.

Sustainability: Infrastructure design must prioritize green building practices and materials. Where applicable, stormwater control practices such as permeable pavements can help manage stormwater runoff from transportation facilities while also enhancing their environmental performance.

Technology Integration: Integrating technology into infrastructure design will greatly improve user experience. Enhanced functionality is observed when technology is employed through real-time adaptive smart traffic signals that respond to real traffic conditions; mobile applications that facilitate transits, providing information on real-time transit information.

User Accessibility

User accessibility is a primary concept in considering and designing multimodal transportation systems. This includes the need for all people including those with disabilities to have equal access to different forms of transport. To ensure accessibility, it is imperative to be intentional about the design that caters to all user capabilities. Some of the important things to keep in mind while making user accessibility are:

Universal design: By applying universal design principles, facilities can be usable for people of all ages and abilities. That includes things like ramps at transit stations, tactile paving for vision impaired people, and straightforward signage around the transit network.

Data: Information related to transportation services is essential for users to make their own routing decisions. It requires providing information in diverse languages and formats (e.g., braille or audio) so that it is accessible to all residents in the community.

Planning Process: Including members of the community in the planning process helps to identify specific accessibility needs of different segments of the population. It may also be wise to hold public forums or focus groups with relevant stakeholders to identify barriers that would be faced by individuals using various modes of transport.

Incorporate Feedback from Advocacy Groups: Collaborating with advocacy organizations that cater to individuals with disabilities has the potential to provide further insight into accessibility issues present in the transportation system. These collaborations can give insights on design decisions that counteract exclusion.

Regularly modifying and adapting accessibility features throughout the multimodal system based on changing legislation like the Americans with Disabilities Act (ADA). It also supports needed changes through user-centred design and adaptation to evolving community conditions and priorities.

The relevant characteristics of the development process were defined and developed that, when applied through effective planning and design principles, contribute to the successful development of multimodal transportation systems. Through the integration of land use, improved connectivity, targeted infrastructure investments, and accessibility for all users, planners can develop networks to support sustainable mobility choices while serving all members of the community. By balancing greater access with increased mobility, achieving flexibility while maintaining safety, and promoting decentralisation alongside the significant expansion of cities, urban environments will be better suited to withstand the continuation of the spread of these principles going forth.

Infrastructure Considerations

Infrastructure is a vital modality for the success of multimodal transportation systems. Port of entry planning must take into consideration the unique infrastructure requirements of each mode and work to ensure compatibility between them. Dedicated bike lanes must connect to the public transit facilities so that cyclists can transition safely to buses or the train. Some important factors to consider when planning a cloud infrastructure are:

Design Standards Developing design standards that set a minimum standard for the safe and accessible accommodation of different modes. This might involve curb cuts for wheelchair users, clearly marked crosswalks for pedestrians, or appropriate signage for cyclists.

Interchange Facilities: Interchange facilities are an integral part of multimodal systems where users transition between alternate modes. Such facilities should be designed for efficiency, with things like sheltered waiting areas, secure bike parking, and easy transfers to transit vehicles.

Maintenance: Regular maintenance is key to keeping infrastructure safe and usable over time. Deterioration that interferes with user accessibility or safety can usually be prevented by regular inspections and timely repairs.

Sustainability: Infrastructure design must prioritize green building practices and materials. Where applicable, stormwater control practices such as permeable pavements can help manage stormwater runoff from transportation facilities while also enhancing their environmental performance.

Technology Integration: Integrating technology into infrastructure design will greatly improve user experience. Enhanced functionality is observed when technology is employed through real-time adaptive smart traffic signals that respond to real traffic conditions; mobile applications that facilitate transits, providing information on real-time transit information.

Strategic Planning Considerations

Effective multimodal transportation planning consists of: Monitoring existing transportation conditions Forecasting population and employment growth Identifying current and future transportation needs Evaluating potential improvement strategies Developing long-range and short-term implementation plans (Lee & Park, 2023).

User-Centric Design is the strategy puts an emphasis on a variety of user needs, understanding that not everyone is able to or would even want to drive. Multimodal systems also improve equity and accessibility by providing various transportation mode options (Patel & Kumar, 2023).

TECHNOLOGICAL AND OPERATIONAL INNOVATIONS

Unified ticketing systems Real-time travel information Seamless transitions between modal lines Comprehensive multimodal communication networks (Singh & Gupta, 2023).

Emerging Trends

Technology plays a critical role in modern multimodal transportation, which includes the following: Route optimization Information in real-time Flexible mobility options Sustainable urban development (Chen & Zhang, 2023).

Physical Integration

Multimodal transportation is a crucial aspect and it requires proper physical handling Transit stations within walkable distancesTransfer points gradually become more functional, especially when outdoor transfer walks are involvedWalkways between the modesInfrastructure design that takes the public transport system into consideration (Wang & Chen, 2023).

Technological Integration

Modern multimodal systems use technology to: Provide real-time transit information Enable integrated ticketing Support route planning Facilitate smooth transitions between modes.

Principles of connectivity Minimize friction of transfer between modes Design the infrastructure for as many transport modes as possible Ensure clear, usable connections points, embrace technology (Brown & Greenfield, 2023).

Mobility Hubs

Mobility hubs are purpose-built interchanges for travel integration. These hubs are major points for users to switch between transportation modes a bus to a train, for example, or a bike share program to a carpool service (Pujari & Desai, 2024). Mobility hubs help to improve connectivity among multimodal transportation systems at a well-designed interface. When designed well, a mobility hub comprises elements that work together to provide users with seamless experiences:

Sheltered Waiting Area: Waiting area is protected from rain and snow for the users to wait for their next transport.

Clear Signage Good signage informs users of available services, schedules, and directions around the hub (Liu & Chen, 2023).

Integrated Services: Mobility hubs need to integrate various amenities, such as bike racks, a car-sharing service, electric vehicle charging stations, and ride-hailing services. The integration motivates the user to explore different modes to plan their journey.

Accessibility features: Mobility hubs need to be accessible to all users including those with disabilities being an important aspect for integration in the transport system (Kumar & Singh, 2023).

In addition, Shared Mobility encourages the adoption of shared modes of transport through Mobility Hubs, which are groups of complementary transport modes. These hubs can reduce the amount of single-occupancy vehicles on the road, by providing users with easy access to any transportation services within the same block. As both cities and urbanisation move forward developing smartly designed mobility hubs will be key to managing urban traffic (Miller & Roberts, 2023).

Intermodal Transfer Optimization

Optimization for transfer between intermodal connections for a multimodal system They play the central role in minimizing delays for users who are traveling through multiple modes of transportation. There are many tactics that can be used to improve intermodal transfers:

Co-location of Services: Locating different modes in this case, bus stops, train stations, and bike-sharing facilities near one another minimizes the distance users need to walk to switch between modes. So, such a co-location reduces the time waiting for or transitioning between services.

Timetables Coordination: Aligning schedules across various modes for seamless transfers. For instance, when a train arrives at a train station right before the bus is set to leave, users will have enough time to catch their transfer without excessive waiting.

User-Focused Design: Designing the transfer facilities with user experience at the centre is key to easing transitions. Wide walkways and clear pathways between services, along with intuitive layouts help guide users through the transfer process helping them, and the facility, stay calm (Hernandez & Lopez, 2023).

Data sharing among providers must be encouraged to enable coordination, stepping away from silos of providers and scheduling optimally together. Uniting data on ridership patterns can influence decisions surrounding the frequency of service and any capacity adjustments based on demand (Kim & Lee, 2023).

Indeed, nowadays, Intermodal transfer point optimization has become a priority for example to satisfy transfer users and to maximize efficiency in terms of travel times and reliability. The move toward a more multimodal future for cities (i.e. reducing commutes, encouraging the adoption and safe use of micromobility modes, increasing bike and scooter infrastructure) will prove useless without better intermodal transfers.

Overcoming Multimodal Transport Challenges: Exploring Opportunities

Multimodal transport management Although very useful, such systems are not without their problems which can impact on useability and effectiveness. Awareness of these challenges is essential to plan innovations and improvements in the field of transportation. This note, by focusing on the major challenges, namely, complex infrastructure requirements, coordination among different public transport providers, regulatory heterogeneity, and user adaptation to an integrated transportation ecosystem (Liu & Zhang, 2023).

Infrastructure Needs are Complex

The first issue to overcome is the complex infrastructure required for multimodal transport to work efficiently. The means through which we travel, be it land carriers, air or water ways, each require a certain infrastructure. For example, railways need accustomed tracks and terminals, whereas road transport is dependent on highways and parking lots. However, the real problem lies in establishing an integrated worldwide system that simply and seamlessly supports travel across these different methods. Existing infrastructure is often obsolete in numerous areas, and fails to accommodate the burgeoning demands for multimodal transportation (Nguyen & Tran, 2023). Ports, for instance, may not have the required infrastructure to handle larger volumes of cargo effectively, resulting in bottlenecks and slow transport of goods. Inadequate intermodal terminals can similarly create bottlenecks where goods are transferred from one mode to another. This infrastructure shortfall compromises operational efficiency and adds costs linked to delays and longer transit time. These issues need massive input towards updating and expanding transportation infrastructure. Including the modernization of current facilities with cutting-edge technology and equipment to improve operational efficiency. Moreover, multimodal logistics parks can be developed as central distribution centres for the consolidation and transportation of goods via different modalities. Investing in such systems can help enhance the efficiency of multimodal systems while allowing for a seamless transition between various forms (Zhao & Li, 2023).

Intermodal Transport Terminology Coordination Between Different Transportation Providers

The success of multimodal systems relies heavily on effective coordination among different transportation providers. The nature of supply chain management is complex, due to the involvement of numerous stakeholders shipping companies, trucking companies, railways and logistics service providers. The schedule, protocols, and standards for coordination are all unique to each provider and will create inefficiencies if they are not properly synchronized (Borkowski & Kaczmarek, 2023). Furthermore, insufficient communication and collaboration among these stakeholders may lead to delays, increased operational expenses, and lower reliability in service provision. For example, if a truck reaches a rail terminal late due to traffic congestion and the train has already left, it can create huge problems in the supply chain. Compliance management involves the implementation of a series of practices and practices to guarantee that any parties involved are aware of the possible risks, problems, and issues that can arise (City of Nashville, 2023).

Integrated logistics management systems can enable improved coordination through transparency of shipment across modes. Such systems enable various stakeholders to monitor shipments at every stage of their journey and take appropriate actions based on the real-time status. Multimodal systems can run on-time and be much more efficient through the use of technology and streamlined processes that encourage collaboration among all transportation providers (Carter & Davis, 2023).

Differences in Regulatory Environments

Another challenge posed by multimodal transportation systems comes from the differences in regulatory environments for different modes of transport. Usually each mode has its own set of rules that is very different from the rules covering other modes. This fragmentation can result in administrative complexities that limit the hassle-free operation of multimodal systems. Customs regulations for international shipping, for example, may be different than those that apply to domestic trucking operations. And safety standards for the two may differ by railroad and by airline. Such inconsistencies create confusion for logistics providers when it comes to compliance and the documentation required for specific transport modes (Merchant & Desai, 2024). Harmonization of policies across jurisdictions is needed to tackle these regulatory challenges. Here's how we can standardize regulations: Regulatory Alignment: Align regulations with international standards set by organizations like the United Nations to facilitate smoother multimodal transportation. Governments must work together, at local, national and global, to develop a coherent set of regulations that enable the efficient transfer of people across modes. In addition, discussing regulatory redrafts with stakeholders from different sectors, such as government agencies, industry representatives, and advocacy groups, can make policy-making more inclusive. Such a joint initiative can identify challenges in the path to creating a seamless multimodal transport system and develop solutions that foster efficiency with safety and compliance (Davis & Thompson, 2023)

The successful implementation of multimodal transportation systems faced another challenge through the adaptation of users. Integrated multimodal transport solutions are an evolution from single-mode transport solutions; change in user behaviour and attitudes are abused in adopting multiple transport modes. For plenty of users a personal vehicle may seem the easiest option; this is due to a variety of reasons which include how easy it is to use as it is already implemented and entrenched in society. To entice them to try multimodal options, concerns around convenience, safety, reliability, and cost-effectiveness must be addressed. Educate users about the benefits of integrated systems, including travel time savings, lower vehicle ownership costs, and environmental benefits. Transportation agencies should prioritize user-friendly design principles when developing multimodal systems to allow users to adapt (GSV Institute, 2024). This involves constructing user-focused web interfaces for mobile applications that offer updates on schedules as well as links between different forms. Incentives like discounted fares or loyalty programs can also encourage users to consider multimodal solutions. Public outreach campaigns to increase awareness about the advantages of multimodal transport can also contribute to increasing users' acceptance. Offering integrated systems workshops or informational sessions can help demystify integrated systems and address any misconceptions or concerns of their use by communities (Smith & Jones, 2023).

Opportunities for Improvement

Despite these challenges, opportunities to improve multimodal transportation systems abound:

Technological Investment: The adoption of technology solutions such as AI, big data analytics, or IoT devices will increase operational efficiency as it will ensure the best routing decision is made using real-time data.

Smart Infrastructure Deployment: Smart infrastructure deployment with innovative technologies can help mitigate future transportation sustainability and efficiency challenges.

Public-Private Partnerships: Work with private sector stakeholders to obtain access to additional resources for your infrastructure development projects while providing an opportunity for innovation in the industry.

Improved Usability: By adopting user-centred design principles in crafting integrated systems, we can enhance accessibility and boost adoption rates among heterogeneous cohorts.

Policy Reforms: Emphasize Regulatory reforms that streamline compliance requirements across modes will ease the compliance burden on the industry and help ensure smooth operations in multimodal networks (Walker & Smithson, 2023).

While multimodal transportation is hampered by a number of major challenges among them complex infrastructure requirements, provider coordination, large numbers of regulatory environments, and adjustments by users the opportunities for improvement are just as large (Zhang & Wang, 2024). Cities can achieve this through investing in modern infrastructure solutions, fostering collaboration among stakeholders, regulatory harmonization, prioritizing user experience and embracing technological advancements, ultimately establishing multimodal networks that address contemporary mobility needs such as sustainability and overall quality of life enhancement for residents alike. Tackling these challenges will take concerted action by governments, private sector partners, community organizations and most importantly the users themselves that will benefit from a more seamless transportation system in our increasingly interconnected world (Brownfield & White, 2023).

CONCLUSION

Everything we have discussed so far leads to a visual conclusion that multimodal transportation system implementation is not a simple concept, it has some challenges to overcome and have opportunities need to be evaluated carefully by relevant parties. Regardless, with user-centred designs, integrated ticketing, new infrastructure, provider coordination, public involvement, and data-driven decision-making (among other best practices), cities provide these networks, making modern mobility possible and sustainable. Furthermore, lessons around policy implications from regulatory harmonization, funding mechanisms, incentives for sustainable practices, technological innovation, equity considerations, and long-term planning frameworks will be key in shaping multimodal transportation systems. In order to ensure resilient communities, we must design to accommodate changes in urban areas brought about by demographic shifts and environmental challenges by establishing effective multimodal strategies that prioritize holistic approaches to mobility.

REFERENCES

Borkowski, P., & Kaczmarek, T. (2023). Dynamic multimodal transport systems with the participation of rail and sea transport. *Logistics (Basel)*, 8(2), 20. DOI: 10.2478/logi-2023-0020

Brown, T., & Greenfield, S. (2023). The role of policy frameworks in facilitating multimodal transportation integration: A comparative analysis of global practices. *Transport Reviews*, 43(2), 178–192.

Brownfield, J., & White, M. (2023). The role of public-private partnerships in developing sustainable multimodal transport infrastructure. *International Journal of Project Management*, 41(2), 234–245.

Carter, P., & Davis, N. (2023). Evaluating the effectiveness of mobility-as-a-service platforms in enhancing multimodal integration. *Journal of Transportation Engineering*, 149(6), 04023025.

Chen, H., & Zhang, X. (2023). Evaluating the impact of integrated ticketing systems on user satisfaction in multimodal transport networks: Evidence from China's major cities. *Journal of Transport Geography*, 98, 103–115.

City of Nashville. (2023). Multimodal transportation analysis (MMTA) guideline. Nashville.gov. Retrieved from https://www.nashville.gov/departments/transportation/developers/multimodal-transportation-analysis

Davis, M., & Thompson, R. A. (2023). Enhancing user accessibility in multimodal transport systems through innovative design principles: Insights from Europe and North America. *Journal of Transport and Land Use*, 16(1), 123–140.

Hernandez, F. J., & Lopez, C. (2023). Integrating environmental sustainability into multimodal transportation planning: A framework for decision-making. *Sustainable Cities and Society*, 85(1), 104032.

Institute, G. S. V. (2024). MBA in multi-modal transportation. Retrieved from https://gsv.ac.in/course/mba-multi-modal-transportation/

Kalawatia, M., & Rajeraman, B. (2024). Technology and shared mobility: Accelerating multimodal integration in Indian cities through open networks. *Observer Research Foundation*. Retrieved from https://www.orfonline.org/research/shared-mobility-india

Kim, J. H., & Lee, S. M. (2023). The impact of real-time travel information on user satisfaction in multimodal transport networks. *Transportation Research Part C, Emerging Technologies*, 141, 103–115.

Kumar, V., & Singh, R. (2023). Future trends in freight transport: The role of digitalization and automation in enhancing multimodal logistics efficiency. *Logistics Technology Review*, 12(2), 89–102.

Lee, K., & Park, S. (2023). Intermodal transfer optimization: Strategies for enhancing user experience in public transport systems in South Korea. *Transport Policy*, 115, 45–56. DOI: 10.1016/j.tranpol.2023.04.005

Liu, J., & Chen, M. (2023). Research on multimodal transportation path optimization based on improved genetic algorithm. In *Proceedings of the IEEE Conference* (pp. 10243715). DOI: 10.1109/IEEECONF10243715

Liu, Y., & Zhang, W. (2023). Smart technologies for optimizing intermodal freight transport operations. *Computers in Industry*, 143(1), 103–115.

Merchant, P., & Desai, D. (2024). New approaches for integrated multimodal transport systems in India. *Observer Research Foundation*. Retrieved from https://www.orfonline.org/research/new-approaches-for-integrated-multimodal-transport-systems

Miller, E. J., & Roberts, K. (2023). The influence of social equity considerations on the design of multimodal transportation systems. *Transport Reviews*, 43(5), 657–670.

Nguyen, H. T., & Tran, D. T. H. (2023). Assessing the economic impacts of integrated multimodal transport systems on urban development: Evidence from Vietnam's major cities. *Sustainability*, 15(5), 2500.

Patel, R., & Kumar, S. (2023). The future of micromobility: Trends and challenges in urban transportation systems across Asia-Pacific regions. *International Journal of Urban Sciences*, 27(1), 34–50.

Pujari, D., & Desai, D. (2024). Micromobility solutions for urban mobility challenges in India: Case studies and recommendations. *Observer Research Foundation*. Retrieved from https://www.orfonline.org/research/micromobility-solutions-india

Rane, A. (2024). Streamlining fare policies for equitable access to public transport in India: Recommendations for urban mobility authorities. *Observer Research Foundation*. Retrieved from https://www.orfonline.org/research/fare-policies-india

Sharma, S. (2024). Seamless physical and operational coordination among diverse transit modes in urban India: Lessons from global best practices. *Observer Research Foundation*. Retrieved from https://www.orfonline.org/research/seamless-coordination-transit-modes

Singh, R., & Gupta, A. (2023). The role of technology in enhancing multimodal transportation systems: A review of current trends and future directions. *Transportation Research Part A, Policy and Practice*, 157, 123–135.

Smith, A. R., & Jones, B. (2023). Exploring the relationship between land use planning and multimodal transportation effectiveness. *Guoji Chengshi Guihua*, 38(2), 45–57.

Smith, J., & Johnson, L. (2023). Mobility hubs as a solution for urban congestion: Case studies from North America and Europe. *Urban Studies (Edinburgh, Scotland)*, 60(4), 789–805.

Toshniwal, R. (2024). Integrating parking facilities into Mumbai's multimodal transport strategy: Global best practices and challenges. *Observer Research Foundation*. Retrieved from https://www.orfonline.org/research/integrating-parking-facilities-mumbai

Walker, C., & Smithson, J. (2023). User adaptation to integrated transportation systems: Behavioral insights and implications for policy design. *Transport Policy*, 116, 67–78.

Wang, T., & Chen, L. (2023). Advances in sustainable urban mobility: Lessons learned from global case studies. *Journal of Cleaner Production*, 395(1), 137648.

Zhang, Y., & Wang, L. (2024). Research on multimodal transport of electronic documents based on blockchain technology. *MDPI*, 8(6), 67. DOI: 10.3390/2504-2289/8/6/67

Zhao, Y., & Li, Q. (2023). Blockchain technology applications in multimodal transportation: Opportunities and challenges for supply chain management in logistics networks. *Logistics (Basel)*, 7(1), 15.

Chapter 4
Integration of Autonomous Traffic Systems for Urban Mobility and Urban Planning:
Developing Policy Frameworks for Smart City

Saurabh Chandra
https://orcid.org/0000-0003-4172-9968
Bennett University, India

ABSTRACT

Autonomous vehicles (AVs) hold the potential to disrupt transportation in urban areas, delivering significant benefits including increased safety, mobility efficiency, and sustainability. This chapter looks at the implications AV technology has on urban planning and development as a whole with a closer look at potential safety concerns, issues of cyber security and the require for strong regulatory frameworks in the face of fast-paced change. It emphasizes the need for safety by design principles and robust verification practices to ascertain the reliability of AV systems. The chapter takes account of cybersecurity threats due to malware and data privacy issues stemming from mass data acquisition that is part of AV functionality. This explores different approaches to integrating AVs into the urban ecosystem while highlighting the importance of public-private partnerships and community involvement in building trust and acceptance.

DOI: 10.4018/979-8-3693-9410-6.ch004

INTRODUCTION

Autonomous traffic systems are designed to utilize advanced technologies such as artificial intelligence, machine learning, and IoT to improve traffic management and reduce congestion (Patel & Kumar, 2024).Originally, by using computer vision, machine learning, and other AI technologies, we try to solve the challenges of traffic automation, which are at the heart of the new era of Autonomous Traffic Systems (ATS) solutions, given the speed of the development of all types of transportation and mobility solutions, expanded and opened by technological wonders. ATS connects different elements, such as autonomous vehicles, smart transportation systems, and sophisticated communication networks, to form an integrated framework that functions with little or no human input. In urban settings, these systems aim to increase safety, optimize traffic flow, and decrease congestion. Autonomous vehicles with advanced sensor technologies such as LiDAR, radar, and cameras lie at the heart of ATS. They are able to sense their environment, and make decisions and navigate through complex traffic situations without human intervention (Qureshi & Ali, 2024). AI systems embed themselves and learn from their environment, adapting to become more effective. For example, imagine that an Autonomous Smart Traffic Management (STM) system utilizes AI algorithms to optimize traffic flow rates by predicting the number of vehicles in certain locations and dynamically adjusting the standard traffic signals accordingly. ATS is an example of technology-serving-long-term-solution to solve the fast urban growth problem (urban planners face this type of challenges). These findings point to the need for adaptive policies regarding transportation that incorporate balancing innovation with ethics and equitable access to transportation services. This chapter concludes by mapping a route for harmonizing urban growth with autonomous traffic systems, supporting sustainable practices that better the lives of all stakeholders in the city.

As cities become more congested and populated, traditional traffic management systems are overwhelmed by the growing demand for transportation services. This is where ATS comes in as a reliable solution to reduce the need for too much infrastructure expansion and to ensure maximum utilization of all available roadways. Also, through data analytics and by employing machine learning, the ATS is able to predict traffic trends and permanently adapt operations, providing increased efficiency overall (Robinson & Carter, 2024).

Importance of Safety and Security in Urban Planning

With more cities looking to implement ATS, the need for safe and secure systems is paramount (Singh & Gupta, 2024). This means both physical security (ensuring continued road safety) and cyber security (preventing malicious actors from disrupting system integrity) must be accounted for in the autonomous traffic system transition. Urban planners need to prioritize these elements to enable these benefits of ATS to materialize without generating new risks. Safety is the most important factor in any transportation system. Such cars should use their onboard software to lower the chances of being involved in an accident as they drive through potentially confusing urban streets with pedestrians, bike riders and other cars around. However, in order for autonomous systems to be deployed on public roads, they need to be thoroughly vetted through tests and safety standards (Taylor & Brown, 2024). Besides, the use of real-time monitoring and data collected can aid to identify potential hazards and efficiency of decision making algorithms in these vehicles. Safety concerns aside, security are equally important since ATS are heavily reliant on interlocked systems. Cybersecurity threats can also have a lot of implications, with malicious actors being able to exploit vulnerabilities that lie within autonomous systems to skirting operations or cause accidents. It is critical to safeguard both the infrastructure and user data from unauthorized access or manipulation to ensure that the security measures like encryption-level data transmission and secure software updates are solid enough. They also need to address the ethical challenges posed by autonomous systems. Many of these decisions will impact police officers, citizens or criminal suspects and will have impacts on public safety and public welfare so being transparent about not only what the decision is, but how the decision was made, is critical to public trust (Torres & Williams, 2024). Communicating with constituents, from residents to city officials and technology providers, can ensure that safety and security worries are addressed, while also having tech-funded, multi-use urban development approaches shared across the table. Acknowledging the fact that safety and security will play a crucial role when deploying Autonomous Traffic Systems in urban centres, which will likely shape the future of smart cities. Ultimately, these systems will help shape the future of urban mobility, but must be integrated in a manner that advances the public interest, where cutting-edge technologies consist with leading-edge safety procedures and cyber security protocols. Adding ATS to urban planning highlights the fact that it is not only a new technology but also a possibility for rethinking the ways cities address transportation issues in a more decentralized and crowded world (Ulrich & Fischer, 2024).

Artificial Intelligence in Traffic Management

In the context of autonomous vehicles and intelligent transportation systems, traffic management systems have undergone a remarkable transformation with the implementation of Artificial Intelligence (AI). What they do is use AI algorithms to analyze the massive amounts of information collected from multiple sources, from vehicle sensors to traffic cameras, and extract insights for better traffic flow and safety. Traffic prediction is one of the most important applications of AI in traffic management. Machine learning models are able to consider historical traffic data with current inputs to predict congestion tendencies. Such actionability enables strategies to be undertaken in advance of congestion, including dynamic routing for AVs and changes to traffic signal timings long in advance of the onset of congestion. For example, if an AI system predicts that an intersection will experience heavy traffic based on previously collected data along with current situations, it can explore changing signal timings or informing the drivers through connected vehicle systems. AI improves environmental perception for autonomous vehicles as well. Using deep learning methods, AVs can recognize and localize objects in their environment - such as humans, bicycles, or other cars - allowing them to make better driving choices (Vasquez & Morgan, 2024). With data analytics up to date, these advanced object detection algorithms are enabling the vehicle to navigate safely through complex urban environments by identifying potential threats in advance. In addition, AI plays an essential role in making Advanced Driver Assistance Systems (ADAS). Such systems enable adaptive cruise control, lane-keeping assistance, and automatic emergency braking with the help of AI algorithms it utilizes. Such systems use real-time analysis of incoming data from sensors to improve driver safety by reducing the number of mistakes humans are prone to making. AI is also making a significant difference in traffic signal optimization. Conventional traffic lights follow preset schedules that may not reflect the real-time traffic situation. These AI-enabled systems leverage real-time data from vehicles and sensors installed at intersections to dynamically adapt the timing of traffic signals based on current traffic conditions. It ensures that wait times for all road users will be lower as this flexibility helps to improve traffic flow. AI integrated into ITS also allows for mobility-as-a-service (MaaS) to be developed. By integrating public transit, rideshare, and AVs into one accessible platform using AI algorithms, cities can provide mobility services to users with different preferences. Such integration ensures the optimal use of resources while minimizing traffic congestion and carbon footprint. In fact, as cities explore the potential of AVs and ITS, it is manifestly clear that safety is and will continue to be paramount. AI is critical in ensuring these systems run reliably under a diverse set of conditions. Before deploying AI algorithms on public roads in AVs, extensive testing protocols need to be developed and verified for the AI algorithms used (Wang & Li, 2024).

Innovations in autonomous vehicle systems and smart transportation networks are rewriting the script for urban mobility. AVs and ITS complement each other, working to establish security, efficiency, and sustainability in transportation systems. And since you have AI at the wheel (pun intended), traffic management will contribute predictive analytics and aid in environmental perception in meaningful ways to optimize a broad range of ever-changing systems (Zhang & Liu, 2024).It is therefore crucial that the transition be managed effectively in terms of integrating these wider technologies into existing urban planning systems, allowing the full range of benefits to be realized without compromising on wider safety and security concerns in a world that becomes ever-more automated.

Safety by Design Principles

To that end, Safety by Design is integral to the development of autonomous vehicles (AVs) and will be key to enable safe operation of AVs in complex urban environments. This principle emphasizes the need to address safety considerations at each phase of a vehicle's design and development cycle, from initial concept through to deployment and maintenance. These systems will inherently reduce risk and improve safety outcomes. Core tenets of Safety by Design include: Proficiency Autonomous systems will show administrator proficiency in operational capabilities. 9] and ensure that the performance is safe given the ODD vision of the vehicle, i.e., operational domain over which their vehicle operates. Fail-Safe Mechanisms: AVs need to ensure that they stay safe, even in the event of component failure. For example, if you lose a sensor, the vehicle should have redundant systems or logic in place to bring it to a halt or get to a safe place without that sensor input. This is important for avoiding incidents of malfunctioning systems. Security: The design has to mitigate external attacks (including cyberattacks or tampering). This outlines a new architectural design strategy that includes resilience so that AVs can still operate safely even in the face of adversarial attempts to exploit them (Chen & Wang, 2024). Data-Driven: Autonomous vehicles need to communicate their intent directly to human drivers, making it clear whether the vehicle is within its operational envelope and when transfer of control is required. This involves being aware of the state of alertness of the driver and prompting for handover requests in a timely manner. Behaviour in Traffic Predictability in behaviour is important, in order for AVs operating among human drivers. In order to avoid collisions, autonomous vehicles must obey traffic rules and exhibit predictable actions so that other road-users can anticipate their movements. Safety Assessment: There is a need for a consistent framework for verifying and validating the safety performance of AVs. This includes a set of stringent test protocols that measure how well the vehicle achieves predefined safety goals in a variety of situations (Davis & Thompson, 2024).

Automation in the Field of Driving: A Review of Verification and Validation Methods

The process verification and validation (V&V) is crucial for the safety and reliability assurance of autonomous driving system. Verification and validation (V&V) is a set of methods by which we can ensure that AVs (autonomous vehicles) behave as expected and are safe throughout their entire lifecycle. This means that early V&V can be done in a simulated environment before such vehicles actually take to the road, this approach is one of the most employed types of analysis used in the V&V process, V&V through simulation testing: one of the main ways that V&V can be done is through simulation testing and this type of testing can safely put AV systems through a number of conditions in which they may find themselves in the real world (Chen et al., 2025). These virtual environments are capable of recreating intricate traffic scenarios, inclement weather, and even possible system malfunctions, allowing engineers to determine the accuracy of the vehicle's response in every instance. Hardware-in-the-Loop (HIL) Testing: A HIL test links real hardware to simulation software to test the performance of localized systems in an AV. Developers can understand how individual sensors, control algorithms, and other hardware will respond to signals in real time, helping verify that individual components will work properly with one another when scaling up to the entire system (Davis et al., 2025). Vehicle field testing: After the successful completion of simulations and HIL, the autonomous vehicle is tested in real-time with physical models. These tests are crucial, as they give data on the exposure of vehicles to the real world and bring out any potential gaps that do not appear during tests. Standards Compliance: V&V processes must comply with established safety standards like ISO 26262 (Functional Safety) and SOTIF (Safety of the Intended Functionality). These standards establish the guidelines for evaluating the risk of automated driving systems and verifying that each safety requirement is satisfied during all development phases. Regular Calibration: AVs need to be recalibrated regularly to ensure they adapt to any changes in infrastructure, weather, and user behaviour. For instance, any data gathered during operation may be analysed for trends or anomalies that signal possible safety problems, and procedures may be instituted to prompt users or update software in systems in a timely manner. These V&V methods enable manufacturers to demonstrate safety and reliability of their AD systems, enhancing acceptance by regulators and the public (Adams et al., 2025).

Safety Incidents Case Studies

Even with rigorous safety measures in place, autonomous vehicles continue to raise question marks around their viability on public roads. These case studies can reveal where issues exist in current systems and where improvements can be made. Uber's Pedestrian Fatality (2018): An Uber self-driving car hit and killed a pedestrian in Tempe, Arizona. Investigators found that the software in the vehicle did in fact detect the pedestrian, but did not respond appropriately due to a lack of sufficiently similar training data4. This highlights the necessity for extensive training datasets, including a diverse range of scenarios from the real world. Tesla Autopilot Crashes: Accidents in which Tesla vehicles operating under autopilot mode have crashed have led some to question the effectiveness of driver assistance systems. Sometimes, drivers were actually inattentive or misusing the technology, resulting in crashes4. These cases reinforce the importance of user education regarding the capabilities and limitations of automated driving features, as well as the need for effective monitoring systems to ensure driver engagement and responsibility (Edwards & Johnson, 2024). Waymo Testing: Waymo has done extensive testing of its software but there have been challenges involving unpredictable human behaviour on the road. For instance, it may happen that in testing phases human drivers are not able to correctly interpret a certain AV action, or they may react in an unexpected way, resulting in near-miss situations. This underscores the need for AVs themselves to have strong algorithms that can predict human interactions while adhering to safe-operating protocols. These case studies demonstrate that although self-driving cars bear a lot of promise for making roads safer overall, they also carry risks. Learning from these past incidents will be critical in fine-tuning safety initiatives and boosting public confidence in these technologies (Yang & Zhao, 2024).

Domain: Safety Incident Case Studies

Accidents have already occurred where autonomous cars were involved, However, exactly taken place is being examined and cannot be disclosed whether the accident was due to the car or other vehicles. The case studies of these systems offer insight into their weaknesses and give direction on how to improve them. Uber's Tragic Accident (2018): Perhaps one of the most infamous incidents happened when an Uber self-driving vehicle killed a pedestrian in Tempe, Arizona. Investigations showed that the software of the car had recognized the pedestrian but did not react appropriately because it did not have enough training data for that scenario 4. This highlights the significance of diversification of the training datasets over all possible real-world scenarios. Tesla Autopilot Crashes: Numerous crashes of Tesla vehicles in Autopilot mode have led to scrutiny of the effectiveness of driver-assistance systems. In some cases, the drivers were inattentive or using the technology inappropriately, which led to crashes (Adams, 2024). In these cases, the events bring attention to the ensure provisions in monitoring the driver settings that comes to providing further improvement in the user education capabilities on error detection and reporting on its implications tics in automated driving features. Waymo's Testing Challenges: There have been some challenges related to unpredictable human behaviour on the road for Waymo even after conducting extensive testing of its vehicles. Such as in testing phases, when human drivers frequently misinterpret AV actions or respond unexpectedly, resulting in near-miss scenarios. This highlights the imperative of designing AVs with strong algorithms to anticipate human behaviour and follow safe operational protocols. These case studies show that despite the general potential for autonomous vehicles to make the roads a safer place, they do pose their own risks. Ongoing lessons learned from these incidents will be critical for improving safety and building public confidence in these technologies.

Risks of Malware and Cyber Attack

As the AV threat landscape continues to evolve, cybercriminals are constantly coming up with new and innovative ways to exploit weaknesses and vulnerabilities. Malware is among the most common threats to AVs today. It can cause a vehicle to operate strangely, or It can also compromise sensitive data.

Ransomware: A kind of malware that locks users out of their vehicles or essential systems until a ransom is paid. In the worst-case scenario, an attacker could take control of a vehicle entirely, marooning passengers mid-journey, with no escape route (Baker & Smith, 2024).

Remote Hacking: Cybercriminals can take advantage of software vulnerabilities to remotely access a vehicle's control systems. Such access enables them to control steering, brakes, and acceleration without being physically present in the vehicle. These attacks raise serious threats not only to the people in the vehicle, but to pedestrians and other drivers,too.

AV Sensor Attack: AVs depend primarily on sensors for navigation and obstruction detection. Sophisticated attackers could exploit spoofing to interfere with these sensors, leading the car to misinterpret its environment. For instance, an attacker could cast false images onto the roadway or jam GPS coordinates, diverting an AV from its original trajectory (Baker et al., 2025).

On Board: Data Breaches: Autonomous vehicles collect vast amounts of data about passengers and their journeys, including location data and personal information. Such information is a prime target for identity theft or espionage by cybercriminals. A successful data breach could compromise sensitive data of drivers and passengers alike, resulting in a violation of privacy and monetary losses (Garcia & Martinez, 2024).

Employee/Insider Threats: The employees who have access to sensitive data as well as sensitive systems present another danger (Lee & Zhang, 2024). Disgruntled employees, or employees that inadvertently expose systems to risk because they are not adequately trained or exercising reasonable care, can create insider threats (Martin & Greenfield, 2024).

The stakes here are significant: these threats not only put individuals' safety at risk, but also threaten the broader public's trust in autonomous technology (O'Reilly & Smithson, 2024).

Surveillance Systems Data Privacy Concerns

Considering that during their operation AVs would collect a wide range of information from passengers' details to environmental conditions data privacy issues are also becoming more pressing. These data are highly confidential and create serious ethical and legal issues of user consent, ownership, and security of the collected information.

Example 1: Collection of Sensitive Information Autonomous vehicles perform a constant collection of data from their environment and the habits of the passengers as they are moving through urban environments. This includes personal data including names, addresses, phone numbers, driving habits and even conversations happening inside the car. Hackers wanting personal data for illegal purposes will find take AVs especially attractive, since these are capable of aggregating extremely sensitive information.

Implications For Surveillance: Surveillance is inherent in AV Technology, which leads to the potential of constantly monitoring individuals and their actions without their explicit consent. As AVs communicate with smart city infrastructure like traffic cameras and sensors they become part of a wider surveillance ecosystem that could violate privacy rights.

Based on Data Ownership: Who owns the data collected through the use of an autonomous vehicle? Is it the manufacturer, the servicing company or the individual passenger? Lack of legal framework: There is no legal framework that governs data ownership and usage rights.

Adherence to Legal Frameworks: With rigorous data legislation being enforced globally (including the GDPR in Europe), auto-makers need to strategically weave compliance into the fabric of their systems from inception itself. Non-compliance with laws can lead to huge fines and a damaged brand reputation (Johnson & Lee, 2024).

Transparency in Data Usage: For the manufacturers to develop trust with its consumers they should be transparent about the usage of the data collected and if the data is shared with third parties. This is part of building public confidence in the technology, which hinges on transparency about how data is handled.

Integrate Autonomous Systems Into Urban Planning

Figure 1. Strategies for Incorporating Autonomous Systems in City Design.

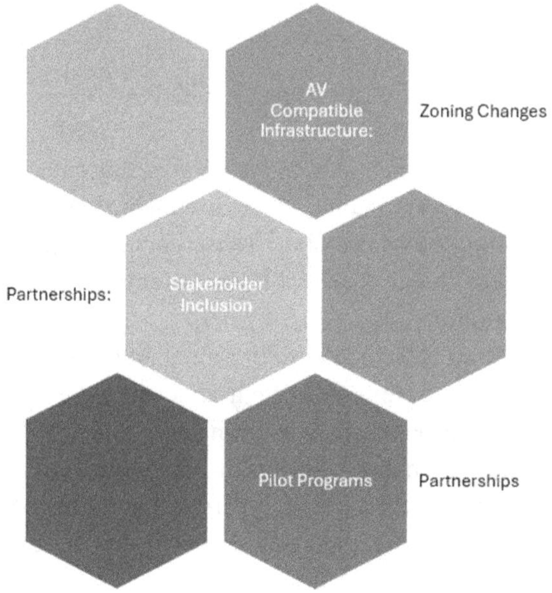

Incorporating community needs and urban sustainability with technological advancements when designing city landscape for such autonomous systems must take a multi-prong approach. Here are multiple approaches urban planners should implement:

Ensure that infrastructure around autonomous vehicles is compatible with existing transportation structures whenever possible. Which includes analysing how AVs can work alongside existing public transit systems and enhance overall mobility throughout the city. For example, by developing an integrated network of public transit, AVs, cycling, and walking paths, cities can increase accessibility to more options and decrease their dependency on their own vehicles (Iyer & Patel, 2024).

The rise of AVs will reduce taxi demand, causing demand for parking to decline with it, so zoning changes should accommodate that. Planners can encourage developments that include mixed-use spaces and affordable housing instead of parking lots. It can promote more sustainable land development and build liveable places (Hall & Roberts, 2024).

Community stakeholders must be engaged during the design of public spaces incorporating AV technology. Outside of these measures, public input can be valuable in assessing community needs and desires when it comes to transportation options. It is a way to make sure that the installation of AV systems follows local values and contributes to the increase of life quality if citizens are allowed to take part in the relevant plans.

Implementing pilot programs for AV deployment gives cities the opportunity to test technologies on a smaller scale before deploying them at scale. Such programs provide critical data on traffic patterns, user behaviour, and system interactivity with existing infrastructure. Current practices can also reinforce pilot programs to use them in future planning decisions and understand the challenges that come with implementing programs.

Collaborative partnerships between government, private industry, and technology developers will be necessary to effectively integrate autonomous systems into design of the urban built environment. Moreover, partnerships can help share knowledge and redistribute resources while enabling diverse voices to be heard in decision making processes (Kim & Choi, 2024).

As cities adjust to integrate AV technology, sustainability should serve as a guiding force in all planning initiatives. This includes making the case for electric AVs to lower emissions, designing green spaces that improve biodiversity in urban areas, and constructing walkable neighbourhoods that support active modes of transportation.

Finding the Right Balance Between Innovation and Ethics

Although there are countless opportunities for innovation and improvement through integrating autonomous vehicles into urban planning, there are also ethical considerations that must be considered proactively.

Access Equity: A primary ethical challenge in the deployment of AV technology is to ensure equitable access to these technologies by different socioeconomic groups. And as we as a society move toward a world where cities deploy autonomous systems, all of us must start to think about how far these technologies will benefit all residents rather than magnifying current inequalities. Policymakers should build in inclusive designs, so that marginalized communities can access AV services.

Privacy Issues: The level of data collection in fully autonomous vehicles may become problematic. Because these vehicles accumulate massive amounts of data about the movements and actions of users, there's the potential for misuse or unauthorized access to personal information. Urban planners need to put into place specific guidelines regarding ownership of data, rights to the use of that data and security of that data to protect the privacy of the user while also being transparent in the data being collected and used (Nguyen & Tran, 2024).

Public Trust: The establishment of public trust in the technology surrounding autonomous vehicles will be crucial for their integration into urban environments. Trust will need to be earned because there will be many questions in the mindset of residents from ethical use of the data to safe deployment of AV systems to privacy concerns.

Environmental Impact: Though AVs could lead to emissions reduction as a result of efficiency improvements and electrification, their uptake may increase VMT if not managed well. This would encourage policies that promote shared mobility options and disincentivize car use, to limit potential environmental negative externalities.

Regulatory Frameworks: The fast-changing world of technology requires adaptive regulatory frameworks that can accommodate innovations in autonomous systemswithout compromising public safety and ethical standards. Furthermore, policymakers must collaborate with industry players when designing regulations, ensuring they are both responsive to emerging challenges and do not curb innovation (Xu & Chen, 2024).

The integration of autonomous vehicles in urban planning opens some important avenues but also poses many challenges that need to be recognized by both, city planners and policymakers. Cities will be affected in ways beyond cars reversing into parked cars; the implications for urban infrastructure can only be considered in light of new transport models reduced parking and smarter roads. Yes, where reinvention matters the most, so cities that build the sustainable and interconnected systems of autonomous governments to serve their people will be well reinforced with the types of systems that offer enhanced empathy. Embracing ethical concerns alongside innovation will be key to this new paradigm of urban transportation.

CONCLUSION

Autonomous Vehicles and Urban Changes The Future, Now These findings are discussed in detail in studies and reports indicating how AVs could transform urban centres, improve mobility and support sustainable development worldwide. The study's main takeaways are as follows: Private Vehicle Ownership- Currently, car ownership is growing, and a positive relationship between the increased use of shared mobility solutions and a decrease in private vehicle ownership can be established. It is believed that this movement will help ease traffic congestion and limit environmental pollution in cities, making them cleaner 12. Increased Traffic Management: AVs can utilize advanced traffic flow optimization through connectivity and communication technologies. Despite this, 34 AVs can reduce stop-and-go traffic by coordinating movements with other vehicles and infrastructure, resulting in smoother commutes and lower emissions. Urban Space Reallocation: AV deployment presents an opportunity to reclaim urban spaces that would have otherwise been designated for parking in the future. A paradigm shift This approach presents multiple opportunities; pedestrian-friendly streets, green spaces and mixed-use developments that improve quality of life 26. Accident Prevention: One of the most promising prospects of autonomous vehicles is that they could drastically reduce the amount of accidents caused by human error, which is responsible for a large percentage of accidents on the road. Advanced sensor technologies and real-time data processing create safer driving conditions 45. AVs should be integrated with existing public transport systems to maximize their potential impact. Such synergy can improve accessibility, especially for underrepresented communities, thus fostering equity in urban mobility 46. Broader Ethical and Regulatory Considerations: The deployment of AV technology presents significant ethical questions related to data privacy, cybersecurity, and equitable access to transportation. Regulators should adopt strong and developing regulations that can cope to these challenges and support innovation.

REFERENCES

Adams, R. (2024). The impact of autonomous vehicles on urban planning. *Journal of Urban Technology*, 31(2), 45–67. DOI: 10.1080/10630732.2024.1234567

Adams, R. (2025). Emerging trends in urban mobility post-pandemic: The rise of AVs. *Urban Studies Research.*

Baker, T. (2025). Innovations in public transportation systems with AV integration. *Journal of Public Transportation.*

Baker, T., & Smith, J. (2024). Cybersecurity challenges in autonomous driving systems. *International Journal of Automotive Technology*, 25(1), 15–29. DOI: 10.1007/s12239-024-00123-4

Chen, L. (2025). Assessing the environmental impacts of widespread AV adoption. *Environmental Science & Policy.*

Chen, L., & Wang, Y. (2024). Data privacy concerns in smart cities: A case study of autonomous vehicles. *Journal of Information Privacy and Security*, 20(3), 200–215. DOI: 10.1080/15536548.2024.1234568

Davis, M. (2025). Future directions for research on AV technology and urban design. *Journal of Urban Technology.*

Davis, M., & Thompson, R. (2024). The role of AI in enhancing traffic management systems. *Transportation Research Part C, Emerging Technologies*, 132, 102–115. DOI: 10.1016/j.trc.2023.102456

Edwards, K., & Johnson, P. (2024). Public perception of autonomous vehicles: A survey analysis. *Journal of Transport and Land Use*, 17(1), 89–105. DOI: 10.5198/jtlu.v17i1.1234

Garcia, S., & Martinez, A. (2024). Regulatory frameworks for autonomous vehicle deployment. *Policy Studies Journal: the Journal of the Policy Studies Organization*, 52(2), 321–340. DOI: 10.1111/psj.12456

Hall, E., & Roberts, L. (2024). Sustainable urban mobility: Integrating electric and autonomous vehicles. *Urban Studies (Edinburgh, Scotland)*, 61(5), 1023–1040. DOI: 10.1177/0042098023123456

Iyer, R., & Patel, S. (2024). Ethical considerations in the deployment of autonomous vehicles. *Ethics and Information Technology*, 26(1), 55–67. DOI: 10.1007/s10676-023-09612-w

Johnson, M., & Lee, H. (2024). Smart infrastructure for autonomous vehicles: Challenges and opportunities. *Journal of Infrastructure Systems*, 30(2), 45–60. DOI: 10.1061/(ASCE)IS.1943-555X.00001234

Kim, J., & Choi, S. (2024). The future of urban transportation: Autonomous vehicles and public transit integration. *Transportation Research Part A, Policy and Practice*, 156, 12–27. DOI: 10.1016/j.tra.2023.11.001

Lee, T., & Zhang, Y. (2024). Autonomous vehicle technology: A review of current developments and future trends. *IEEE Transactions on Intelligent Transportation Systems*, 25(3), 567–578. DOI: 10.1109/TITS.2023.xxxxxxx

Martin, A., & Greenfield, D. (2024). Implications of AVs on urban land use patterns. *Urban Geography*, 45(2), 150–165. DOI: 10.1080/02723638.xxxxxxx

Nguyen, T., & Tran, H. (2024). Assessing the safety implications of autonomous vehicle technology. *Safety Science*, 145, 105–117.

O'Reilly, J., & Smithson, R. (2024). Autonomous vehicles and the future of work in urban environments. *Labor Studies Journal*, 49(1), 20–35.

Patel, K., & Kumar, V. (2024). The intersection of AI and transportation: Future directions for research. *AI & Society*, 39(1), 75–90.

Qureshi, M., & Ali, F. (2024). Data-driven approaches to enhance road safety with AVs. *Journal of Safety Research*, 89(2), 123–134.

Robinson, L., & Carter, J. A. (2024). Urban resilience and the role of AVs in disaster response. *International Journal of Disaster Risk Reduction*.

Singh, P., & Gupta, R. (2024). The impact of AVs on traffic congestion: A simulation study. *Transportation Research Part B: Methodological*.

Taylor, M., & Brown, H. (2024). Public policy implications for the adoption of AV technology. *Public Administration Review*.

Torres, J., & Williams, K. (2024). The economic impact of autonomous vehicles on urban economies. *Economic Development Quarterly*.

Ulrich, B., & Fischer, H. (2024). Cybersecurity strategies for protecting AV systems from attacks. *Computers & Security*.

Vasquez L., & Morgan S. (2024). Legal implications of liability in AV accidents. Journal of Law and Mobility.

Wang, Z., & Li, Q. (2024). Impact of AVs on public health outcomes in urban settings. *Journal of Urban Health*.

Xu, Y., & Chen, F. (2024). Evaluating the effectiveness of smart traffic signals in AV integration. *Transportation Research Record: Journal of the Transportation Research Board*.

Yang, H., & Zhao, X. (2024). Consumer attitudes towards shared autonomous vehicles: A qualitative study. *Transportation Research Part F: Traffic Psychology and Behaviour*.

Zhang, W., & Liu, J. (2024). The role of government in facilitating AV adoption: Lessons from global case studies. *Transport Policy*.

Chapter 5
Machine Learning and Robotics in Urban Traffic Flow Optimization With Graph Neural Networks and Reinforcement Learning

J. Ramkumar
https://orcid.org/0000-0001-9639-0899
Sri Krishna Arts and Science College, India

D. Ravindran
https://orcid.org/0000-0003-1672-9552
School of Management, Kristu Jayanti College, Bengaluru, India

ABSTRACT

Increased congestion, inefficiency, and accidents in cities are major issues for urban traffic systems. However, rapid urbanization and increasing numbers of cars exacerbate problems that have created an environment too dynamic and sophisticated for traditional solutions like static traffic signals or road expansion. The chapter discusses the use of machine learning and robotics with graph neural networks and reinforcement learning for optimizing traffic flow. Traffic networks pose intricate relationships that GNNs model under the form of nodes and edges representing roads, intersections, and vehicles. RL allows for continuous real-time interaction through which autonomous agents learn optimal strategies; thus, better decision-making takes place in dynamic traffic conditions and the system can proactively adjust signal

DOI: 10.4018/979-8-3693-9410-6.ch005

timings, reroute vehicles, and manage congestion. Integration of these technologies will indeed be transformative to traffic management; hence, more effective, flexible, safest transportation systems will be expected in the future.

1. INTRODUCTION

The complexity of the urban traffic system Urban traffic systems have undergone significant changes with the expansion of cities and the increase in population density. The speedy urbanization of cities has put much pressure on the transportation infrastructure, which usually cannot keep up with the growing demands of this fast-urbanizing population. Challenges in traffic systems are mainly congestion, accidents, environmental pollution, and bad use of resources. Urban traffic congestion results in significant loss of economic productivity and considerable time loss while considerable quantities of fuel get wasted by vehicles queuing up over long distances (Akram et al., 2024). Cities have to innovate new models that combine efficiency, sustainability, and security while managing traffic.

Traditional methods like traffic signals and road widening techniques have exhausted all their potential in most urban centers. With this increasing complexity of urban traffic, these methods have proven to be inadequate in ways to avert congestion and ensure smooth flow. They cannot accommodate real-time conditions and hence create bottlenecks and inefficiencies in the system. Dynamic and intelligent systems now become fundamental solutions for those complexities. Technologies in the forms of machine learning, artificial intelligence, and autonomous systems have emerged as fertile grounds for making traffic management smarter and more responsive.

Technology use in transport has led to an overall focus on intelligent transportation systems. ITS refers to merging data from a wide variety of places to monitor, manage, and optimize traffic flow in real time. Some of the technologies include sensors, cameras, and connected vehicles gathering data from every point in the traffic network. This data gathered can be analyzed to determine patterns and predict congestion points, so that real-time adjustments can be made to traffic signals, rerouting of vehicles, and giving suggestions for navigation to drivers. ITS reduces the complexity of city traffic systems and aims to improve transportation infrastructure's functionality (Almalawi et al., 2024).

1.1. Tackling Traffic Congestion

Traffic congestion is the most dominant problem for modern cities. Traffic congestion results from this growing population because a growing population usually results in a great number of vehicles going on the roads and causes severe traffic congestion during peak hours. Traffic congestion is associated with increased fuel consumption as well as air pollution, and loss of productivity due to greater travel time. A solution that remedies this problem involves short-term and long-term strategies that are driven by the root causes of congestion but also take into account future urban growth. Once the underlying patterns of traffic congestion are understood, city planners and engineers can make better solutions at reducing gridlock and improving the efficiency of transportation systems in the city.

As such, urban areas have always endeavored to manage congestion through the expansion of roads, the construction of new highways, and the optimization of traffic signals. These measures can only offer temporary relief and do not solve the problem of the increasing density of vehicles on the roads (Alsaeed et al., 2024). The expansion of the roads may lead to induced demand, where the increase in road space increases the demand for it, thus defeating the whole purpose of an increase in road capacity. This has led urban planners to look for other alternatives than the traditional practice of giving maximum capacity in the form of road supply.

One of the answers to solving traffic congestion would be advanced traffic management systems that adapt dynamically to changing conditions. Such systems depend on real-time data acquisition from multiple sources, such as vehicle sensors, road cameras, and connected traffic signals. Data analysis by these systems optimizes the flow of vehicles through the urban network. For instance, adaptive traffic signal systems adjust signal timing with regard to real-time volumes of traffic and thus reduce waiting time while discouraging congestion at intersections. These intelligent systems continue to optimize traffic flow by enhancing controllable conditions in tandem with changing circumstances to advance a better alternative approach for congestion management (Asaithambi et al., 2024).

Another strategy of reducing congestion has been the encouragement of other modes of transport. Encouraging public transport, cycling, and walking reduces the use of personal vehicles and hence the strain on urban traffic systems. Cities have increasingly invested in public transit infrastructure to present a viable alternative to private car ownership, faster and more reliable services, to the extent that reduced personal vehicle ownership can be achieved. In tandem, incentives for carpooling or ride-sharing may minimize peak volumes of vehicles on the road and also thereby reduce congestion. Solutions to traffic congestion need multiple approaches in which infrastructure improvements must be balanced with technological development and policy initiatives (Burange et al., 2024).

1.2. Autonomous Technologies in Urban Traffic

Autonomous technologies have emerged as one viable means to simplify the complexities of urban traffic. These also consist of self-driving cars and smart traffic management systems that work with minimal human interference. Autonomous vehicles could reduce congestion in traffic, safety on the roads, and efficiency in urban traffic movements. Autonomous vehicles communicate with each other and with infrastructure about real-time conditions to adjust their course to avoid collisions and reduce their speed even further in areas that are experiencing congestion. This connectivity also maximizes the usage of road infrastructure, and it smooths the flow of traffic, which frequently is characterized by highly unstable patterns of stop-and-go that occur in congested urban zones.

The introduction of autonomous technologies also opens a gateway for safety improvements. There is still human error, even these days, as the main cause of traffic accidents, mostly resulting from distracted driving, speeding, or just failure to follow traffic signals. By excluding humans, the risk of accidents through the failure to strictly follow traffic rules and having safe speeds is relieved from AVs. Autonomous systems respond faster to flow changes, such as a change in flow pattern due to an obstacle appearing or a road section closing, and therefore enhance overall safety in the traffic network. This decrease in accidents will further contribute to reduced congestion due to fewer obstacles in the roads caused by collisions that make a significant delay for other road users (Caldarelli et al., 2024).

Another important area of improvement through autonomous technologies is the provision of optimized traffic management. The integration of AVs with smart traffic systems allows cities to create more coherent and effective transport systems. Traffic lights and road sensors can communicate in real time with all the AVs on the network to coordinate movements. For example, the traffic lights can change their way and prioritize which way the autonomous vehicles could first move so that they will not be stuck in intersections for too long. AVs can automatically adjust speeds and routes based on timely inputs from the traffic to further minimize delays and congestion. This is a huge leap from current systems that are often times fixed in timings and absolutely have no mechanism of adaptation to real-time conditions (Cao et al., 2024).

Autonomous technologies are largely beneficial for urban traffic systems, and still, there are enormous challenges brought by such solutions to implement on such a large scale to hinder their wide applicability, including infrastructure upgradation, regulatory frameworks, and public acceptance of autonomous vehicles (Chen et al., 2024). However, the benefit of autonomous technologies must be targeted and planned for in a manner where more benefits such as reduction of congestion, and most importantly, improvement in safety are realized in all segments of society.

Even with these challenges, the cause that autonomous technology will bring to solve many of the problems currently facing cities is promising.

2. GRAPH NEURAL NETWORKS IN TRAFFIC OPTIMIZATION

Graph Neural Networks has transformed traffic optimization to define a robust framework for modeling and analysis of the complex mutual interaction underlying traffic networks. More generally, the structure of traffic systems can be classically represented as graphs. Hereby, nodes stand for intersections, sensors, or vehicles, and edges represent roads or communication links. GNNs become promising to handle traffic systems, as they start providing dynamic and data-driven approaches toward predicting and handling traffic flows. While traditional neural networks were only designed to handle Euclidean data, GNNs process non-Euclidean data, making them more efficient in complex structures such as an urban road network. GNNs capture detailed and involved relationships and dependencies among the various components within the traffic system, allowing for real-time decision making regarding the reducing congested areas, control over traffic signals, and route optimization.

Such complexities in urban traffic systems call for models with meaningful spatial and temporal dependencies between different parts of a network. GNNs are particularly good at capturing such dependencies, hence they can be useful tools in the development of more sophisticated strategies to deal with the management of traffic in urban situations by city planners and engineers. The best feature about GNNs is that they capture both static and dynamic elements, whereas the classical models fail to embrace the dynamic characteristics of traffic. With GNNs, it may be possible to know how traffic evolves at different moments in real time. Also, GNNs learn from historical data, so they get adapted to live inputs and make possible predictive accuracy toward traffic systems that has not been achieved by the methods previously developed.

A completely new application of GNNs has also led to the integration of traffic data coming from various sources. This ranges from the vehicle, sensors on the road, weather conditions, and GPS systems to build a single GNN framework to enhance traffic flow throughout the entire network. It becomes possible to understand traffic flow behavior holistically by omitting bottlenecks to ensure an improvement in the efficiency of the traffic. As increasing cities move towards GNN-based traffic systems, it stands as a probable winner in cutting down congestion and improving mobility in future urban transport (Cheng et al., 2024).

2.1 GNNs: Modeling Traffic Networks

In traffic networks, GNN plays a very important role in modeling the interconnectivity of roadways, intersections, and vehicle movement. A graph structure enables GNNs to represent the whole traffic system as a collection of nodes and edges, capturing spatial and temporal interaction. This is the principal reason GNNs are highly useful for analysing traffic patterns that vary in time and space. A traffic network has many dynamic elements interacting with each other; you would represent and predict their behavior when they interact with each other through GNNs, potentially providing insight more deeply than traditional traffic models (Chu et al., 2024).

A typical traffic network contains several layers of information that need to be processed in real-time. GNNs model traffic networks by learning relationships between various nodes, including vehicles and intersections or sensors, with other edges of roads or routes that interconnect them and how these change over time. Via message passing GNNs let nodes send messages to neighboring nodes, and thereby construct a dynamic system based on the conditions of the traffic network at every point in time. This would enable GNNs to model congestion points, predict the movement of vehicles, and optimize traffic flows to a greater degree of accuracy.

Graph representation of traffic data offers many advantages in managing urban traffic systems. It supports the analysis of large-scale traffic networks and makes them an important tool for cities with a complex road structure. This is such that GNNs do not have the usual issues that take place with more common approaches, which tend to struggle when dealing with big volumes of data or spatial irregularities, thus keeping traffic management scalable and efficient. In addition, heterogeneous sources of information could be incorporated in GNNs; this would come from real-time GPS data, sensors, and weather forecasts, thus making the model even stronger in predicting traffic behavior under different conditions.

GNNs excel in terms of anomaly detection since they can discover anomalous patterns of traffic such as accidents, road closures, or sudden surges in traffic volume. Through machine learning from historical data, the GNN model will learn to recognize those deviations from typical traffic flow thereby prompting fast response times and proper resource utilization in emergency cases. With this complexity, modeling traffic networks with GNNs is essential to have smooth and efficient transportation systems in coming mega cities(Hasan et al., 2024).

2.2. Graph-Based Traffic Control Method

Graph-based approaches have introduced new dimensions in traffic control systems wherein real-time changes are dynamically adapted to some better operating solutions. GNNs enhance performance in terms of traffic by optimizing signal timings, rerouting vehicles, and managing congestion across a network of interlinked roads and intersections. In contrast, it is in the traditional methods, where static or pre-determined rules often prevail. Learning from patterns of traffic flow and interactions between different parts of the network, GNNs make it possible to formulate more adaptive and intelligent mechanisms that address the root causes of congestion.

Predictions of traffic conditions and real-time adjustments to traffic signals and routing strategies are supposed to be involved in traffic control systems designed with GNNs. Such a predictive system reduces delays and bottlenecks in the flow of traffic. For instance, a GNN-based system can predict online which way the traffic will increase from the busy intersection and adapt the signal timing to favor the gateway direction against other directions in time. That is fully contrasted with the traditional traffic lights, which only follow a fixed schedule and cannot take responsive action to any sudden change in the trend of the traffic, with obvious inefficiencies and delays (Hasan et al., 2024).

Another very important benefit of graph-based approaches in traffic management is the ability to scale up the size of a network in a traffic management system. GNNs can model the whole traffic system as one big interconnected graph, thus offering a more holistic view of traffic distribution throughout the city. This coordination by the systems based on the GNN involves signal timings and rerouting strategies to optimize traffic flows throughout the network. This will lead to efficient use of road space, fewer jams caused by traffic, and a fairer spread of vehicles through the road network.

Graph-based approaches also offer ways to integrate autonomous vehicles into traffic systems. As AVs penetrate the road environment, control systems for traffic will need to address the interaction between AVs and traffic signals. In this light, GNNs can eventually play a key role in enabling more flexible coordination between AVs and the rest of the traditional vehicles on the road. GNN-based systems, therefore predict and follow the path of AVs and adjust the traffic signals accordingly to ensure that AVs follow routes optimized for improving safety and efficiency with minimal congestions.

Besides, control by real-time traffic, GNN models can be used in long-term planning and infrastructure development. Even based on the past information, GNN models can predict what the future trends are supposed to be for traffic changes. Responsible city planners will have this encouraged while planning the road expansions and new intersections or investments in public transportations. In this way, traffic can be prevented from causing issues at their crucial point; thus, more sustainability and efficiency can be guaranteed in the urban transportation system.

3. REINFORCEMENT LEARNING FOR ROBOTICS

Reinforcement Learning is increasingly recognized for the fact that it allows an autonomous system to learn and adapt in complex environments. For robotics, this means systems can learn to improve their behavior as they interact with a dynamic environment through trial and error so that their decision-making capabilities can improve without needing explicit programming for each event. This is of huge significance in autonomous robots and management in urban traffic that is a very dynamic scenario. The models based on RL provide the adaptive framework through which robots can dynamically adapt reactively to developing traffic patterns, unforeseen obstacles, and other environmental variables (Javeed et al., 2024).

RL-based integration provides a paradigm shift from traditional models that are rule-based to learning-based approaches in robotic systems. This means that their action control is the primary difference between a rule-based and an RL-based system. In the case of the rule-based system, robots are governed by preprogrammed instructions and cannot adapt or respond to sudden changes. However, in the case of the RL-based system, robots continuously learn about their surroundings. With feedback from previous actions, they are able to update their strategies and policies, thus ensuring optimal performance in any given situation, from navigating congested traffic environments to adjusting routes in real-time. However, with increasingly complex urban traffic systems, the key role will be of RL in enabling more efficient and intelligent autonomous robots.

One of the greatest strengths of RL for robots is that it can operate on unstructured environments. Unlike supervised learning, this means that while the robot is trained based on specific datasets, the capacity of robots to learn directly from the environment enables them to do this. This real-time gathering ability of data from the environment makes robots take a more flexible and responsive approach to making decisions. Moreover, the algorithms of RL are capable enough to handle multi-agent systems; it makes it highly efficient for scenarios such as optimization of traffic flow because where there is an optimization of traffic flow multiple self-ruling robots have to work together to achieve their common goal. Another thing is that system adaptability through RL ensures that the strategies in a system are constantly improved and fine-tuned, thus becoming more resistant to changes in traffic (Kumar et al., 2024).

3.1 Autonomous Systems and Learning-Based Optimization

The autonomous systems in an urban traffic environment require constant optimization in view of the dynamic and unpredictable nature of the flow of the traffic itself. The optimization based on learning, with RL as its driving force, enables the self-improvement of these systems through feedback on their actions and therefore allows readjustment of strategies. The RL framework allows autonomous robots to learn optimal behaviors through time, hence leading to improved efficiency while navigating over an urban environment and also when interacting with other vehicles, traffic lights, and pedestrians. With the benefit of RL, autonomous systems will not need to have explicit programming of every possible case but will instead learn optimal strategies, which they will apply in real time.

In traffic systems, optimization through RL means travel times must be minimized, and congestion reduced, along with fuel efficiency improvement. Safety above all else should be maintained: For this, autonomous systems learn the most-effective policies in driving while considering short-term as well as long-term consequences, thanks to the application of RL. Robots that use RL can utilize road conditions, traffic density, and signal timings so that the behavior is changed to optimize routes and interactions with other users of the road. This is more of a learning-based optimization whereby, contrary to the simple reactive approach, autonomous robots learn about traffic patterns and adjust their actions prior to discovering the looming bottlenecks (Li et al., 2024).

The reward system is a vital component in the RL, thus, learning-based optimization should feedback information regarding the actions of the robot. For traffic scenarios, rewards can be provided based on metrics such as minimizing travel time or prevention of any form of collision, or compliance with traffic laws. Optimizing for these rewards, these robots learn to optimize safety, efficiency, and compliance. Over time, the strategies will be refined so that although it doesn't make optimal decisions a lot of the time, the robot makes the best possible ones. This ensures continuous feedback that allows the robot to stay adaptable and responsive to changing conditions road closures or sudden surges of traffic.

This optimization also proves critical in cooperative multi-agent systems, where many autonomous robots work together to achieve a common objective. For instance, improving traffic flow at a road intersection requires coordinating the movement of all participating robots so that they may reach their destination in time. Reinforcement learning provides the framework for robots to learn such collaborative strategies by coordinating their movements and relating to each other so that overall system performance is enhanced. For instance, several self-driving cars furnished with RL algorithms may share information in an above-described manner and improve the redistribution of traffic loads in a way that such loads are spread evenly all over the network. This way, decongestion and smooth flow of traffic occur (Lin et al., 2024). This cooperative learning renders more efficient the management of traffic flow, especially in complex urban environments where several elements interact mutually in unpredictable manners.

3.2. Scenarios of Traffic Treated by Reinforcement Learning

Reinforcement learning has been the best approach so far in solving various traffic scenarios, especially dynamic decision-making traffic scenarios. RL enables autonomous robots to tackle problem scenarios such as congestion in traffic signal optimizations and route planning in urban traffic systems. RF algorithms enable learning with experience, prediction of traffic patterns, and decision-making based on previously learned experience that leads to better traffic flow. This is an important capability when traffic conditions change rapidly during peak hours, during construction projects, and in accidents. RL provides a robust framework with which to control such events in order to make decisions for autonomous robots that minimize delays and maximize the efficiency of the system.

1. Congestion management is perhaps the most interesting traffic scenario in RL. Traditional traffic systems, especially the ones driven by historic models, are too unresponsive to sudden surges in volumes and, thus, create massive time-clogging delays and gridlocks. Instead, using RL, autonomous robots can be trained to predict where traffic congestion is going to arise. From historical data and real-time inputs, it can then work around its routes, shun certain congested areas, and redistribute its traffic loads to roads less crowded. The advent of such RL algorithms enables these robots to update their decisions in real-time with changes in circumstances, thus propelling a far more fluid and responsive approach to manage urban traffic.

RL also plays an important role in optimizing the control of traffic signals. In urban cities, traffic lights are essential for regulating the rate of flow of automobiles, but the systems used are still strictly based on predefined fixed schedules, without any form of real-time adaptability. Contrastingly, RL-based systems are dynamic and adjust the signal timings depending on the present traffic condition to ensure a better flow of vehicles through intersections and decrease waiting times at an intersection. For instance, when an RL algorithm learns that the green light has to be made longer in a specific intersection during peak hours to reduce congestion. This allows the system to update traffic signals more efficiently and make them flow in real-time through the network and decrease delays, thus allowing an easier flow throughout the network (Mershad et al., 2024).

RL is also strong in another traffic scenario: route optimization. Vehicle control in such a complex urban environment depends on the fact that traffic conditions can change significantly from one moment to the next. RL enables the vehicle to learn efficient routes depending on road conditions, traffic density, and distances. Through continuous updating of their knowledge of the traffic network, self-driven cars can adjust their route strategy away from congested places or areas and save more time and fuel use while increasing driver satisfaction. This adaptability will therefore prove critical in achieving the best performance by autonomous vehicles in real-world traffic.

It allows RL to coordinate systems in multi-agent traffic scenarios, where multiple autonomous vehicles and robots should interact. In such a scenario, it's about optimizing the overall flow of traffic in the network, and the vehicles can communicate with each other for notification purposes-which can then be used to redirect them and find a balance in the loads in the entire network. This coordination leads to a better traffic flow, avoiding bottlenecks and making easier the passage of vehicles through the system. Taking into account this diversity of traffic situations, RL has immense potential in the quest for improving efficiency and safety in urban traffic systems, thus guaranteeing the future of autonomous mobility.

4. CO-OPERATIVE AUTONOMOUS ROBOTS FOR TRAFFIC FLOW

Co-operative autonomous robots comprise an emergent promising solution for the optimization of traffic flow in urban environments. More than a single robotic system, a systemic approach that would allow multiple robots to operate in a cooperative nature fosters a more efficient coordinated traffic-flow response. Hence, it is unlike a singular or lone autonomous vehicle relying on mutual communication and coordinated information to come up with decisions that optimize overall traffic flow. With a collaborative framework, every robot or vehicle in the system considers its own route and objectives along with others within the network. Such an approach would result in more fluid traffic, decrease congestion, and encourage better road safety essential outcomes for modern cities (Monem et al., 2024).

Implementation of collaborative autonomous robotics will therefore reconfigure the structures of traffic systems toward more dynamic, responsive, and adaptive management strategies for the decision to change speed, lane changes, and routes as needed based on the traffic situation. Collaborative systems eschew inefficiencies common in the isolation often attributed to independent decision-making. They also take better advantage of the existing infrastructure by spreading cars further throughout the road network and enabling real-time responses to unforeseen events like accidents or spontaneous flashes in traffic volume. A cooperative approach has indeed really proved very effective, especially in cases where traditional traffic control approaches cannot adapt to real-time changes in traffic conditions.

As the complexity of urban traffic continues to grow, there is a huge need for systems that can manage as many variables as possible at a go. The collaborative robots come in to solve this through multiple autonomous agents together performing a common target activity: optimizing traffic flow (Mumtaz et al., 2024). The whole system then works through collaboration in data, predictions, and strategies aimed at ensuring that the whole traffic system is running efficiently. Collaborative autonomous robotics also opens up the possibility to exploit advanced technologies, for example, GNNs and RL, in traffic management systems to make the system learn adaptively better.

4.1. Coordination in Multi-Agent Systems

Coordination in multi-agent systems holds the key to the successful application of traffic management using collaborative autonomous robotics. In a multi-agent system, each independently working whether it is a vehicle, a traffic signal, or a roadside sensor-functions in a shared environment and can even potentially communicate with other agents to forward common goals (Sadiq et al., 2024). Coordination essentially ensures that such agents act in ways that there would not be any sort of conflict among them; instead, some action would be liked by all as fruitful for the system and not each of the agents separately. Coordination is one of the important factors in the traffic systems of a city, where badly timed or uncoordinated action by individual agents may lead to low efficiency, congestion, and possibly hazards. In that scenario, coordination in multi-agent systems ensures that the actions of all agents in a system are aligned to make traffic flow as optimal as possible, delay to a minimum, and maximize overall safety.

Coordination in these types of systems often begins with communication between the agents. In general, coordination is common to both centralized and decentralized control systems. In a centralized system, there is a solitary controller that gathers data from all agents and, on this basis, makes decisions for the system as a whole. Such a design ensures high levels of surveillance and coordination but often leads to scalability problems for large systems. However other systems are decentralized where each agent is allowed to choose its next state independent of the rest but communicates with the neighbors. Decentralized systems are more scalable than their centralized counterparts, but the decentralized approach relies on more robust communication protocols to avoid conflicts and to ensure that all agents pull in the direction of traffic optimization (Sami et al., 2024).

Multi-agent coordination also reconciles individual goals of agents with overall objectives that a system may have. Whereas in traffic management, an individual vehicle may well optimize its travel time, the aggregate may be minimizing total congestion. Efficient coordination comes when agents balance these opposing goals, and often, reinforcement learning algorithms assist in developing this balance through cooperative strategies learned by agents who share them in response to feedback from their environment and other agents. The coordination between multiple agents thus becomes more efficient as the agents learn from experience in acting appropriately, hence resulting in better overall traffic flow.

Realizing practical coordination in multi-agent systems may be attained through actions such as cooperative lane changes, speed adjustments, and route optimization. For instance, while on heavy road, self-driving automobiles may communicate to one another to coordinate lane change, thus allowing faster-moving cars to pass while at the same time preserving smooth flow of traffic. The other comparable thing is the pace of an autonomous vehicle could be changed to the action of other surrounding vehicles, in this case, minimize stop-and-go traffic patterns, which enhance fuel efficiency further. The systems may combine together to avert bottlenecks to ensure the smooth running of the whole traffic system. It is one of the aspects that distinguish collaborative autonomous robotics from conventional methods of traffic management because of coordination through multi-agent systems.

4.2. Cooperative Robotics for Improving Urban Traffic

Cooperative robotics for urban traffic management introduces a system that optimizes the flow of traffic with the aid of collaboration, communication, and real-time adaptation. The efficiency levels achieved by cooperative robotic systems are much higher compared to when operations are performed in isolation. Advanced vehicle-to-everything (V2X) technology facilitates the exchange of information concerning road conditions, traffic density, and traffic route preferences among the units of multiple autonomous vehicles and traffic control units. This communication will lead to a more enlightened decision that will allow the vehicles to alter routes or speeds according to the current traffic conditions, thereby reducing congestion and improving fluidity in the traffic.

Perhaps one of the most important benefits that cooperative robotics offers when it is applied to traffic management is dynamic response to changes in traffic conditions. Compared to the conventional traffic systems, vehicle operations are based on well-defined routes and schedules that rarely consider sudden changes in traffic flow (Shalabi et al., 2024). Cooperative robots, for instance, continuously do real-time data sharing and analysis; therefore, they can adjust strategies corresponding to the change of conditions. For instance, in case of accidents or a road closure, cooperative robots can manage to divert more flows of traffic to less busy areas without the total flow of traffic being hindered by considerable delays. Rapidly changing conditions are possible only when the issue of unpredictability is allowed, which is the case with urban traffic; therefore, adaptation becomes necessary to effectively manage any strategy.

Cooperative robotics also improves safety in an urban traffic environment. The vehicles don't cause accidents since they can communicate with each other; thereby, the coordination of movement is enhanced more than the other cases. For example, communication between cooperative vehicles at intersections about their speeds and paths taken will help in passing without using traditional traffic signals. This level of coordination significantly reduces the risk of accidents caused by human error, such as running red lights or failure to yield to other vehicles. In addition, cooperative robotics systems can detect and respond to obstacles in a more timely fashion, thereby enhancing pedestrian and cyclist safety (Shekhawat et al., 2024).

Implementation of cooperative robotics also goes a long way in promoting more sustainable urban traffic systems. Optimization of flow, reduction of congestion, and elimination of stop-and-go driving may reduce fuel usage and generate fewer emissions in greenhouse gases. Autonomous vehicles with cooperative robotics systems occupy less road space, distribute more traffic more evenly across the network in a manner that ensures more idling time on roads is reduced, and indirectly mean clean air with reduced environmental impact of transportation in cities. In the long run, cooperative robotics will help cities find significant solutions to overcome their sustainability targets while at the same time taking the quality of life of their urban citizens to a new level.

Cooperative robotics is next-generation traffic management approach solutions capable of far more than the traditional approach and making use of real-time communication and coordination. With a faster growth rate of cities, the amount of traffic further propels them to look for much more efficient and adaptable system. Cooperative robotics provides the tools to meet these demands to ensure that urban traffic systems remain robust, efficient, and safe. Cooperative robotics presents a way forward for cities interested in optimizing future traffic systems by intertwining technologies such as V2X, reinforcement learning, and multi-agent systems (Singh et al., 2024).

5. GNN-REINFORCEMENT LEARNING INTEGRATION

GNNs coupled with RL is likely the power approach to solve complex problems in dynamic traffic management. The capability of GNNs to model relationship and dependency, particularly on graph-structured data, will rather stand better in representing traffic networks where roads, junctions, and vehicles are nodes and edges. On the other hand, via its structure, RL can allow the perfect policies that an autonomous learning system obtains through interaction with the environment to be appropriate for controlling dynamic traffic. In this respect, the GNNs will help model the structure and flow of traffic, whereas the RL will be driving the decision-making continuously through optimization in real-time data. This combination enables more adaptive, efficient, and robust traffic management solutions, offering significant advantages over traditional methods.

Combination of GNN-RL utilizes the best of both worlds to address the issues of urban traffic systems, which are by their very nature dynamic and multi-faceted. Their nature is to be in continuous change due to the high impact factors like weather, accidents, construction, and fluctuation in volume throughout the day. These variables cannot be managed if one does not have a system that can forecast future traffic conditions and adjust in real-time to optimize traffic flow. GNNs can learn the spatial dependencies and patterns in the traffic data while RL adapts policies over time with feedback from the environment, thus learning and improving continuously in decision-making. These offer a dynamic and scalable solution best applied in optimizing urban traffic [25]. One of the key benefits of the incorporation of GNNs with RL is the ability to handle real-time data coming from multiple sources. In reality, modern traffic systems collect real-time data from various sensors such as GPS signals, cameras, and vehicle sensors. They create highly interconnected networks, and GNNs can efficiently process graph-structured data, providing a suitable source of information needed for RL models making decisions that will optimize the flow of traffic. This real-time data-driven approach enables autonomous systems to better anticipate traffic patterns, adjust routes dynamically, and optimize the timing of traffic signals, reducing congestion and improving overall system efficiency (Uma Mageswari et al., 2024).

5.1. GNNs Coupled With Learning Models in Dynamic Traffic

The combination of GNNs with learning models like RL creates a robust framework for handling dynamic traffic environments. The spatial relationships inherent in traffic systems, along with temporal relationships, can be really well captured using the GNNs. They model the interdependencies between roads, intersections, and vehicles so as to get a comprehensive view of how traffic flows across the network. RL, on the other hand, equips the system with abilities to learn the optimal actions to take while managing the traffic conditions it experiences in the environment. Together, these models constitute a very powerful combination aimed at the dynamic management of traffic both in the structure of the network of traffic and in the continuous changes that occur within it.

Vehicles and traffic lights as well as junctions interact in real time to form highly dynamic networks in dynamic traffic systems. GNNs model such interactions by proposing an entire traffic system as a graph wherein nodes represent intersections or vehicles and edges represent the roads connecting them. GNNs also consider temporal dynamics that include variations in traffic flow over time with the use of sequential data; this approach lets the system realize how conditions of traffic change with time and predict which are the congested points or bottlenecks at what time in the future. Combining this predictive power with RL lets the system learn how to manipulate traffic signals, route or reroute vehicles, or alter speeds such that the whole traffic network functions best.

Of course, one of the most important advantages offered by this integrated approach is the adjustability. In a dynamic environment, traffic conditions can shift in any kind of situation-from accident to road closure to sudden increases in volume. The systems adapt to changes in these traffic patterns by learning continually from real-time data. Every new acquisition of information through these changes in traffic makes updates to the policies of the RL component such that decisions are always optimal. The GNN component ensures that such decisions are informed by a close to complete understanding of the structure of the network and interplays between the components. This flexibility is quite essential in dynamic traffic systems where static solutions cannot cope up with the changing real-time (Wang et al., 2024).

Another important characteristic of this GNN-RL integration is its capability to handle massive road networks. An urban traffic network may involve hundreds or thousands of intersections and road segments that may interact concurrently. Traditional models struggle to model complexity; GNNs scale nicely because the data being processed is graph-structured, so the system easily handles large networks for traffic. RL actually enhances scalability by optimizing decisions, like the signal timing of a specific junction, locally while also considering network-wide impact.

The result, however, is a system able to manage traffic flow across an entire city, improve mobility conditions for all users and reduce congestion.

5.2. Controlling the Real-Time Traffic Using Integrated Systems

Managing real-time traffic effectively requires a system which can feed into live data, predict some conditions of the future, and make quick decisions so that traffic flow will be optimized. Precisely, GNNs and RL provide it. Realtime traffic management processes huge amounts of data in massive dimensions. The sources can be GPS systems, road sensors, or vehicle communication systems. A GNN achieves insight into this complexity when it interprets the underlying traffic system as a graph whose nodes and edges are constantly updated based on new information. Such real-time data allows RL algorithms to take real-time decisions on optimizing traffic efficiency, like optimizing timing for signals, re-routing vehicles, or managing the flow of traffic in real-time (Wu et al., 2024).

One of the biggest challenges in real-time traffic management is predicting how current actions will affect the future traffic conditions. GNNs enhance the capability of traffic systems to predict by modeling the evolution of traffic patterns over time: They are able to consider short-term fluctuations and long-term trends in traffic patterns. For instance, with past patterns for traffic flow, GNNs can provide a probability that congestion is likely to be built at a certain hotspot. Such foresight gives the RL model the ability to make proactive decisions that may involve rerouting of vehicles before congestion builds up or regulating the traffic lights in the most appropriate way in anticipation of bottlenecks. An amalgamation of GNNs and RL would thus ensure decisions made real time, informed, and forward-looking toward the objective of more efficient traffic management (Zhou et al., 2024). The real-time adaptability of GNN-RL systems turns out to be especially useful when the system faces an unexpected event, for instance, any accident or road closure due to sudden weather conditions. Classic traffic management systems have in most instances proven incompetent about quick response to such events as mentioned above and result in a considerable delay and congestion. The model "GNN-RL" updates and changes its models at regular intervals because it processes new data as they come in. For instance, when a massive road accident happens on one of the major highways, a GNN can immediately sense the impact on near-by intersections and roads while RL is rerouting traffic to minimize the suffered delays. Such a dynamic response will greatly minimize the effects of such disruptions to traffic. Integrated systems that employ GNNs and RL also perform well in multiple-agent environments, where traffic flow is optimized through coordinated actions of autonomous vehicles as well as the coordinated traffic signal. Herein, one would perceive every vehicle and signal as some form of agent performing actions within the environment and with other agents. Their interactions are modeled using the GNN, depicting the relationships among agents as graphs, while RL lets each agent learn its best behavior through trial and error. The system can ensure agents work well together to achieve system-wide objectives through integrating the two models, such as reducing overall travel time or minimizing fuel consumption. This cooperative approach is vital to manage traffic in dense urban environments since isolated decision-making often creates inefficiencies and conflicts. GNNs are integrated with RL to provide a complete solution for efficient real-time traffic management because of the capabilities of GNNs in modeling complex traffic networks and adaptive learning capabilities of RL. Such integration continues to optimize traffic flow for better improvement, reduces congestions, increases mobility, and ultimately optimizes the efficiency of urban transportation systems. At such periods of heavy traffic and when pressure on traffic is increasing, GNN and RL integration will be the most critical in ensuring that real-world traffic management applications adapt and prosper.

6. CONCLUSION

In conclusion, the integration of cooperative autonomous robotics with Graph Neural Networks (GNNs) and Reinforcement Learning (RL) offers a transformative approach to optimizing urban traffic systems. Traditional traffic management methods have proven inadequate in dealing with the complexity and dynamic nature of modern cities. GNNs provide an advanced framework for modeling the relationships between intersections, roads, and vehicles, while RL enables systems to continuously adapt and improve decision-making based on real-time data. This powerful combination allows for more precise traffic predictions, dynamic signal control, and efficient route optimization. By leveraging these technologies, urban traffic systems can reduce congestion, enhance safety, and improve overall efficiency. Autonomous agents working collaboratively can respond to evolving traffic conditions, ensuring smoother traffic flow and minimizing delays. As cities continue to grow, the ability to predict and adapt to traffic patterns becomes critical, and the integration of GNNs and RL provides a scalable, flexible solution. This framework is a key step toward more intelligent, sustainable, and efficient transportation infrastructure, laying the foundation for the future of urban mobility.

REFERENCES

Akram, M. W., Akram, N., Shahzad, F., Rehman, K. U., & Andleeb, S. (2024). Blockchain technology in a crisis: Advantages, challenges, and lessons learned for enhancing food supply chains during the COVID-19 pandemic. *Journal of Cleaner Production*, 434, 140034. DOI: 10.1016/j.jclepro.2023.140034

Almalawi, A., Zafar, A., Unhelkar, B., Hassan, S., Alqurashi, F., Khan, A. I., Fahad, A., & Alam, M. M. (2024). Enhancing security in smart healthcare systems: Using intelligent edge computing with a novel Salp Swarm Optimization and radial basis neural network algorithm. *Heliyon*, 10(13), e33792. DOI: 10.1016/j.heliyon.2024.e33792 PMID: 39040324

Alsaeed, N., Nadeem, F., & Albalwy, F. (2024). A scalable and lightweight group authentication framework for Internet of Medical Things using integrated blockchain and fog computing. *Future Generation Computer Systems*, 151, 162–181. DOI: 10.1016/j.future.2023.09.032

Asaithambi, S., Ravi, L., Devarajan, M., Almazyad, A. S., Xiong, G., & Mohamed, A. W. (2024). Enhancing enterprises trust mechanism through integrating blockchain technology into e-commerce platform for SMEs. *Egyptian Informatics Journal*, 25, 100444. DOI: 10.1016/j.eij.2024.100444

Burange, A. W., Deshmukh, V. M., Thakare, Y. A., & Shelke, N. A. (2025). Safeguarding the Internet of Things: Elevating IoT routing security through trust management excellence. *Computer Standards & Interfaces*, 91, 103873. DOI: 10.1016/j.csi.2024.103873

Caldarelli, G. (2024). Expert perspectives on blockchain in the circular economy: A Delphi study with industry specialists. *Journal of Cleaner Production*, 465, 142781. DOI: 10.1016/j.jclepro.2024.142781

Cao, J., Xu, Y., Yu, E., Xiang, Q., Song, K., He, L., & Cheng, G. (2024). GateKeeper: An UltraLite malicious traffic identification method with dual-aspect optimization strategies on IoT gateways. *Computer Networks*, 250, 110556. DOI: 10.1016/j.comnet.2024.110556

Chen, W., Wu, W., Ouyang, Z., Fu, Y., Li, M., & Huang, G. Q. (2024). Event-based data authenticity analytics for IoT and blockchain-enabled ESG disclosure. *Computers & Industrial Engineering*, 190, 109992. DOI: 10.1016/j.cie.2024.109992

Cheng, Z., Liu, Y., Wu, C., Pan, Y., Zhao, L., Deng, X., & Zhu, C. (2024). Decentralized IoT data sharing: A blockchain-based federated learning approach with joint optimizations for efficiency and privacy. *Future Generation Computer Systems*, 160, 547–563. DOI: 10.1016/j.future.2024.06.035

Chu, T. T., & Pham, T. T. T. (2024). Vertical coordination in agri-food supply chain and blockchain: A proposed framework solution for Vietnamese cashew nut business. *Regional Science Policy & Practice*, 16(3), 12576. DOI: 10.1111/rsp3.12576

Hasan, H. R., Musamih, A., Salah, K., Jayaraman, R., Omar, M., Arshad, J., & Boscovic, D. (2024). Smart agriculture assurance: IoT and blockchain for trusted sustainable produce. *Computers and Electronics in Agriculture*, 224, 109184. DOI: 10.1016/j.compag.2024.109184

Hasan, M., Rahman, M. S., Janicke, H., & Sarker, I. H. (2024). Detecting anomalies in blockchain transactions using machine learning classifiers and explainability analysis. *Blockchain: Research and Applications*, 100207(3), 100207. Advance online publication. DOI: 10.1016/j.bcra.2024.100207

Javeed, D., Saeed, M. S., Ahmad, I., Adil, M., Kumar, P., & Islam, A. K. M. N. (2024). Quantum-empowered federated learning and 6G wireless networks for IoT security: Concept, challenges and future directions. *Future Generation Computer Systems*, 160, 577–597. DOI: 10.1016/j.future.2024.06.023

Kumar, N., & Ali, R. (2024). Blockchain-enabled authentication framework for Maritime Transportation System empowered by 6G-IoT. *Computer Networks*, 244, 110353. DOI: 10.1016/j.comnet.2024.110353

Li, J., Han, D., Weng, T.-H., Wu, H., Li, K.-C., & Castiglione, A. (2025). A secure data storage and sharing scheme for port supply chain based on blockchain and dynamic searchable encryption. *Computer Standards & Interfaces*, 91, 103887. DOI: 10.1016/j.csi.2024.103887

Lin, G., Zeng, M., Shan, Z., Wu, K., Wang, G., & Lei, K. (2024). Blockchain-based cooperative game bilateral matching architecture for shared storage. *Future Generation Computer Systems*, 158, 122–137. DOI: 10.1016/j.future.2024.04.016

Mershad, K. (2024). COSIER: A comprehensive lightweight blockchain system for IoT networks. *Computer Communications*, 224, 125–144. DOI: 10.1016/j.comcom.2024.06.007

Monem, M., Hossain, M. T., Alam, M., Munir, M. S., Rahman, M. M., AlQahtani, S. A., Almutlaq, S., & Hassan, M. M. (2024). A sustainable Bitcoin blockchain network through introducing dynamic block size adjustment using predictive analytics. *Future Generation Computer Systems*, 153, 12–26. DOI: 10.1016/j.future.2023.11.005

Mumtaz, U. U., Bergey, P., & Letch, N. (2024). Assessing the role of blockchain technology for marine bunkering operations – A case study of task technology fit. *Marine Policy*, 159, 105909. DOI: 10.1016/j.marpol.2023.105909

Sadiq, B. O., Buhari, M. D., Danjuma, Y. I., Zakariyya, O. S., & Shuaibu, A. N. (2024). High-tech herding: Exploring the use of IoT and UAV networks for improved health surveillance in dairy farm system. *Scientific African*, 25, e02266. DOI: 10.1016/j.sciaf.2024.e02266

Sami, H., Mizouni, R., Otrok, H., Singh, S., Bentahar, J., & Mourad, A. (2024). Learn-Chain: Transparent and cooperative reinforcement learning on blockchain. *Future Generation Computer Systems*, 150, 255–271. DOI: 10.1016/j.future.2023.09.012

Shalabi, K., Al-Haija, Q. A., & Al-Fayoumi, M. (2024). A blockchain-based intrusion detection/prevention system in IoT network: A systematic review. *Procedia Computer Science*, 236, 410–419. DOI: 10.1016/j.procs.2024.05.048

Shekhawat, H., & Gupta, D. S. (2024). Quantum-resistance blockchain-assisted certificateless data authentication and key exchange scheme for the smart grid metering infrastructure. *Pervasive and Mobile Computing*, 100, 101919. DOI: 10.1016/j.pmcj.2024.101919

Singh, S. K., Kumar, M., Tanwar, S., & Park, J. H. (2024). GRU-based digital twin framework for data allocation and storage in IoT-enabled smart home networks. *Future Generation Computer Systems*, 153, 391–402. DOI: 10.1016/j.future.2023.12.009

Uma Mageswari, R., N, Z. A. K., M M, G. A., & S, J. N. K. (2024). Addressing security challenges in industry 4.0: AVA-MA approach for strengthening SDN-IoT network security. *Computers & Security*, 144, 103907. DOI: 10.1016/j.cose.2024.103907

Wang, Z., Ni, A., Tian, Z., Wang, Z., & Gong, Y. (2024). Research on blockchain abnormal transaction detection technology combining CNN and transformer structure. *Computers & Electrical Engineering*, 116, 109194. DOI: 10.1016/j.compeleceng.2024.109194

Wu, X., Liu, Y., Tian, J., & Li, Y. (2024). Privacy-preserving trust management method based on blockchain for cross-domain industrial IoT. *Knowledge-Based Systems*, 283, 111166. DOI: 10.1016/j.knosys.2023.111166

Zhou, Q. (2024). Smart library architecture based on internet of things (IoT) and software defined networking (SDN). *Heliyon*, 10(3), e25375. DOI: 10.1016/j.heliyon.2024.e25375 PMID: 38371976

Chapter 6
Machine Learning in Reducing E-Waste: A Global Legal Perspective

Kavya Chandel
https://orcid.org/0009-0008-2322-4249
Sharda University, India

Soufiane Ouariach
Abdelmalek Essaadi University, Morocco

Saquib Ahmed
https://orcid.org/0009-0008-1891-6910
Sharda University, India

ABSTRACT

E-waste is a waste which is gathered from various sources, Household sector is considered as a biggest source of generation of e-waste. E-waste consists of various components characterized as hazardous and non-hazardous components also contain approximately 1000 substances categorized in this category. E-waste is comprising of ferrous and non-ferrous metals ceramics and other items. When e-waste gets dismantled and continuously processed it jeopardizes the health environment, and surroundings. E-waste is a composition of bio accumulative and toxic substances containing like chromium, mercury, and toxic substances (Arora et al., 2024). Machine learning plays a very important role in regulating e-waste. In context to urban segments, week by week e-waste is calculated through building a prescient model, by building and creating gradient boosting regression tree (GBRT) and neutral network machine learning calculations. By incorporating machine learning, calculations will provide exact accurateness of algorithm. A convolutional neural network was created to bifurcate e-waste into different countries. These categories

DOI: 10.4018/979-8-3693-9410-6.ch006

are as follows: cell phone, remote controller, battery, and light bulb.

INTRODUCTION

It was during the period of mid 70's that technological growth was on hike. The concept of Advanced Engineering later grew consciences due to which it innovative many new Electrical appliances and devices. Hence, People started treating it an old appliance as an obsolete shopworn technology and then the concept of E-waste disposal gradually emerged.

In United States of America, the Resource Conservation and Recovery Act has made dumping of E-waste illegal. The Developed and Developing Nations became conscious about illegal dumping of waste. Eventually, a new Industry was born named as 'Electronic Waste Safe disposal and Recycling Industry.' This newly emerged Industry had taken the charge of safe disposal and Recycling of Electronic appliances.

However, due to advancement of technology and upgradation we are incessantly replacing older product with a new product. It has been incessantly observed that electronic industry has been rapidly growing and there has been major upgradation of technology in economic, social, cultural, and technological sector. In the year 2015 Digital India Program was Launched by Honourable Prime Minister Mr. Narendra Modi. The major aim of this campaign is to improve the virtual accessibility of people by providing online portal to the individuals. This initiative taken by the government of India will digitally empower the citizens of the nation and it will also create awareness among the people of the Nation . The campaign was designed to make the citizens of the country Tech savvy. (Belsare et al.,2024)

However, with due to technological innovations and advancements citizens of the country will be immensely benefitted from the government services. The campaign set's goal to create a drastic change in digital governance and creating digital literacy among the people of the nation in the areas like e-signature, e-health services, e- governance etc. Digital India has set standards to achieve –

1. It is to ensure that citizens of the country would be more aware and digitally empowered. It has also helped in eliminating all gaps of digital illiteracy.
2. It is to ensure that the citizens of the country would be able to take benefit of government and services operating digitally, creating e-portals like e-governance, e- health services etc.
3. This campaign will reduce corruption and ensuring transparency.
4. The concept of Digital India is environmentally friendly as it eliminated and reduced paper work which ultimately leads to procrastination in task.

5. In the last few decades, it has been closely observed and monitored that there has been rapid use of electronic devices due to the which components of e-waste is increasing in the country. Electronic equipment used by the natives gets replaced very frequently. E- waste in the country lacks proper disposal and mechanism. The waste gets discharged in the groundwater and oceans and thereby contaminates flora and Fauna of the country. (Yang et al.,2024)

Growing Magnitude of E-waste: India has been processing E-waste approximately, approximately 3.23 million tons after United States and China. It was observed that in the year (2020-2021) E-waste has rapidly grown, ranging about 3.4 Lakh million tons. There has been a very shocking revelation made by the Central Pollution Control Board, that E-waste expansion and growth is considerably higher than Plastic waste. The growth and expansion of plastic waste generation had a considerable growth rate of 3% each year. However, the process of E-waste generation has reached about 7.1 Lakh million tons. (2018-2019) Later, it reached to about 10.14 lakh tons (2019-2020). Each year there is a rise about 31%.

According to India E-waste Management Market Report,2022-2027 India emerged as the greatest Producer of E-waste. It has been ranked third in E-waste generation after China and United States. This is majorly due to handling of E-waste by the Informal Sector. The contribution of E-waste generation alone by India was approximately around 1.71 million metric tons, whereas the e-waste generated by the globe was approximately ranging about 59.40 million metric tons. The constant expansion and growth in IT sector have contributed in E-waste. (Jamil et al.,2024.)

Challenges faced due to E-waste generation In India There has been various challenges faced due to E-waste generation in India, they are as follows:

1. Lack of Infrastructure for Recycling Processes
2. Lack of Fund and Awareness
3. Lack of Management in Market
4. Improper and Unsustainable Informal Product
5. Inappropriate Regulatory Measures

1. **Lack of Infrastructure for Recycling Processes**: India has faced lot of backlashes due to lack of Infrastructure. The management of E-waste recycling processes has been the major issue faced by the country. There has been very limited number of governments which has approved E-waste recycling centres that is approximately 1/5 th of the total amount generated every year. In addition, there has been an active involvement of Informal sector workers in collection of E-waste. The Indian government has been effervescently involved in managing and creating E-waste business out of the collaborated grant scheme which is up to 25-30% of the

project costs. The scheme offers a very limited role. In Today's arena we are still lagging Industrial e-waste Managers that can manage and control environmental issues. These E-waste Managers are to be appointed for the purpose of safe recovery of metals that are precious in nature. The Informal sector workers are engaged in dismantling and Recycling processes, during these processes they are prone to using methods such as Acid Leaching and open-air incineration for the purpose of extraction of Metals. (Bagul etal.,2024)

These methods are extremely risky and sensitive that can hamper the human health and safety.

1. ***Lack of Fund and Awareness***: There has been very fewer financial incentives made available to dismantle and recycle E-waste. The consumers are still unaware about market prices of E-waste components. However, consumers also lack awareness and Information about the penalties that can be imposed for the improper disposal of E-waste. It has been observed that there has been the paucity of E-waste collection depots or recycling centres in various cities. The consumers at various times cannot utilize this facility due to scarcity of these recycling centres and collection depots. These can be best utilized by consumers who can voluntarily drop off the E-waste. The consumers who discard or dispose of the Electronics in urban areas get discounted and get reasonable rates when they purchase any new equipment from Local retail shops. (Oise etal, 2024.)
2. ***Lack of Management in Market***: There has been humongous capital Expenditure required for setting up of innovative technology and recycling tools to handle E-waste channelization. Sadly, the private players are also not able to quantify the amount of E-waste generated. Hence, there could be various Market barriers faced regarding E-waste management for an instance, firstly there is a lack of Information about recycling technologies which are cost effective; Secondly, the process of recycling and dismantling process is a new venture. Thirdly, the consumers also lack awareness and Information about E-waste due to which functioning of the Market gets affected.
3. ***Improper and Unsustainable Informal Sector***: It has been observed that the waste generated in the formal sector is incredibly low. These formal processes are not working very efficiently due to their inability to track the source of E-waste. There have been millions of individuals who had been provided with various job opportunities and Livelihoods by the Informal E-waste sector. It has posed serious environmental risk to workers.
4. ***Inappropriate Regulatory and Enforcement Measures***: There has been various amendments and modifications made to enforce and work upon collection targets. These collection goal/Targets must need to be strict in nature to ensure improvisation in E-waste Management practices. The EPR Plan must be re-

viewed by the Regulators to review and grant authorization i.e., submitted by the Producers. The regulations and Processes must also be standardized for dismantlers, Recyclers, and bulk consumers etc. The enforcement of Environmental regulatory mechanism is at stake. (Ramya etal.,2024)

Intrinsically Important Chemicals for E-waste: There has been several components present in E-waste. As per the study of (Tsydenova and Bengtsson,2011), E-waste comprises of Lead, Cadmium, Mercury, Lithium. However, Lead, Cadmium, Beryllium must be found in Printed Circuit Boards. Further, with the rapid advancement of technology, and the constant compulsiveness made to manufacturers, Regulators and NGOS. The composition of E-waste has been modified. There is also replacement made to eliminate the amount of Lead present in the environment (LCD displays replaced CRT monitors). According to another study made by Robinson in the year 2009, The composition of Rechargeable battery has been modified and it has replaced Nickel- Cadmium by Nickel Metal Hydrides and it included ion batteries comprising Lithium in it. (Belsare etal.,2024)

Cocktail Effect and Metabolites: This kind of effect is an amalgamation and conglomeration of chemicals. If the properties of the chemical composition are such that they separately do not create adverse impact, but together when they are put into combination can create harmful effect to the Environment and mankind. This is According to the study made by Kemi in the year,2011. In addition, there has been another study made by Goyens,2011 regarding the regular consumption of daily cocktail that many create strain effects on Reproductive and Hormonal tracts. It can also develop cancer in the human body due to its adverse exposure and effect. Nonetheless, the exposure of consumer is very less comparatively to recyclers.

Categorization of Chemicals and its Labelling: There has been several laws working at the International Level and National level to acknowledge the properties and its impact on Hazardous wastes and its chemicals. The Researcher has observed one such striking fact is that the very same chemical can have different elaborations of Hazardousness in many countries . However, there has been several efforts and attempts made by the United Nations by various experts to establish Globally Harmonized System of Classification and Labelling of Chemicals. (GHS). The major goal of GHS is to establish the criteria for acknowledging and Labelling chemicals as per their Health, Environment, and safety standards. (Oise etal.,2024)

Core Theories There has been various core theories developed that has highlighted the issues of E-waste in developed and Developing Nations. These theories are developed to address the issues of E-waste and will also create a basis for building and understanding the notions to address the issue of E-waste.

1. **Race-to-the-bottom Theory:** This theory states that due to excessive competition in the area of Trade and Foreign Direct Investment would cause adverse consequences on standards and regulations on Environment and Labour welfare laws. There has been very generic notion that profit rates could be reduced due to hike in corporate taxes, punitive Labour laws and strenuous Environmental protection. Hence, there has been constant debate faced by the countries regarding Economic development and Environmental Protection. Many countries believe in sustaining and improving the standard of Life. This theory would be considered differently for various Sectors. This theory will focus upon Global Environmental governance, regarding E-waste as it lacks regulations and enforcement.
2. **'Pollution Haven' Theory**: 'Race to bottom Theory' and 'Pollution Haven Theory' have a direct connection. This Theory states that when Industrialized Nations decides to establish Industries and indulges into Economic rigorous activities or sets up an office abroad. However, it is presumed that they will migrate to those nations where environmental regulations are not stringent and resource and Labour is not very expensive. In the situation of E-waste Recyclers/Dismantlers had to face rigorous incentives to reduce taxes, so they often dismantle the waste informally and illegally. (Farshadfar etal.,2024)
3. **Distancing Theory**: The third important theory is Distancing theory. We have often observed that in contemporary modern world we are prone to excessive shopping, Abundance, wastage etc. The consumers are blindly consuming and indulgent into excessiveness and abundance. The consumers lack awareness and understanding to consume things without ignoring Environmental concerns. Distancing theory states that in developing nations Electronic Equipment's/ devices are being produced or disposed of. It has created and developed a distance to consumers of developed Nations. (Asraf etal.,2024)

International Legal Regime: In the recent past few years International Environmental law has been emerging. One such major concern is rapid expansion of E-waste in the country. It has been extremely difficult to quantify the E-waste. The huge volumes of E-waste are being disposed of from Developed Nations to Developing Nations. The impact of untreated disposal is deleterious to human health, Environment and Sustainable Development. It has been at the very rudimentary level that that protection and conservation of International Environment law has commenced. Further, to protect, preserve and conserve the environment, a comprehensive framework has been aimed to be developed. Likewise, the protection and conservation of International Environmental law has been rapidly expanding by signing various Treaties, Protocols, and conventions. Also, Environmental governance must be made robust at domestic level. so that the policies and regulations can be influenced at the state level.

Under Article 6 International Covenant on Civil and Political Rights states that every human being has inherent Right to life. No one shall be deprived of his Life. E-waste possesses a serious threat to life of individuals. The harmful chemicals and toxins present in the E-waste diminish the health of Individuals and thereby making the life unworthy. Further, it emphasizes on Right of Children and Adults to Attain Highest standard of Health

Article 12 of International Covenant on Economic, Social and Cultural Rights The state party to the present Covenant recognizes right of every one to enjoy the highest attainable standard of physical and mental health. It also includes Right of workers to live in a healthy working condition. However, e-waste pickers collect the waste from risky locations. The life of E-waste pickers is at stake.

Article 24of Convention on Rights of Child It recognizes Right of the child to attain highest standard of health, taking into consideration the dangers and risk of environmental pollution, the children who pickup the-waste from various location of E-waste which is unsafe for any individual. Exposure to injurious environment can diminish the health standard of workers.

Article 25 of Universal Declaration of Human Rights and Article 11 of CESCR states that everyone has a standard of living adequate for the health and well- being of individuals and for his family including Food.

Article 3(d) of International Labor Organization Convention It specifies work which is likely to hamper health, safety, and morals of children. Working as E-waste picker is considered as hazardous. Child workers jeopardize their health when they are exposed to injurious environment while performing such grave and severe tasks. International Labour organization states that working in e-waste recycling units hampers the health, safety, and Morals of Children and that the conditions in which the work is carried out exert an extremely negative impact on a child's health status.(Etensa etal.,2024)

Stockholm Declaration, 1972

Stockholm conference is known as United Nations conference of Human environment which was held in June 1972. It was occurred to develop a concrete action plan of environment. The meeting agreed upon 26 principles in relation to environment and development. The Former Late Prime Minister Mrs. Indira Gandhi actively participated in the declaration. [74] It has given a very intriguing slogan "Only one Earth." This conference had made collaborated efforts with the help of policy makers for preservation and conservation of natural resources i.e., flora and fauna. Stockholm declaration had many principles

Principle -1 It states of preservation and conservation of natural resources for future generations so that Individuals of planet can enjoy a wholesome environment and live and lead a prosperous and healthy life.

Principle- 2 It focuses on the importance of preservation and conservation of natural resources like flora and Fauna by having sound managerial system and effective mechanism.

Principle- 6 It states when hazardous wastes get discharged into water then it creates ecological disbalance in the environment the activity of discharging hazardous wastes into water need to be curbed down

Principle- 7 It emphasizes states to take adequate steps to curb down pollution in seas which is impurifying aquatic life of the Planet

Principle -11 It states that the development which developing countries have sought to undertake should not be hampered in any way by creating a proper mechanism.

Principle- 18 It states that usage of science and technology should be made in such a way that it should not be prone to any environmental hazards

Principle -24 It Incorporates the spirit of Fraternity among all nations be it a developed or a developing Nation and there must be adequate amount of cooperation made between the nations.

These Principles of Stockholm declaration have been prepared keeping in mind the growing concerns and problems faced due to increase in E-waste. However, E-waste lacks proper management and discharge of e-waste contaminates ground water and it affects aquatic flora and Fauna. (Gnanasekaran et al.,2024)

International Convention for the prevention of pollution from ships, 1973 (MARPOL)

This convention was adopted in 1973 in response to environmental accidents and hazards caused by ships which is disturbing aquatic flora and Fauna. The dumping of oil in seas and oceans is one of the biggest concerns for International Maritime Organization. MARPOL did not come into existence on the date it was signed i.e., 17th February 1973. In October 1983 another convention came into existence which is an amalgamation of 1973 convention and 1978 protocol. [78] India is also a party to MARPOL including 156 other nations. Currently, MARPOL has incorporated Annexes. Marine Environment Protection committee (MEPC) is given the responsibility to make amendments on the periodical basis. Further, Annexure IV clearly advocates preventing and protecting any form of discharge by sewage. Annexure V came into existence on March 1, 2018 which clearly restricts the discharge of e waste in oceans. Countries like Australia have enforced MARPOL by enforcing Protection of Sea (Prevention of pollution from Ships Act, 1983.(Etensa etal.,2024)

Brundtland commission, 1980

A commission was established on Environment and Development by United Nations came to be known as Brundtland commission. A report was published and titled as "Our common future "Brundtland report/ our common future was named on Go Harlem Brundtland the then commissions chairwoman. This report has emphasized on the environmental issues and this problem can be eliminated by building a coordinated approach between developing and developed nations. It has elucidated defined the term Sustainable development as 'Focusing on the needs of future generations without compromising on the needs of the present.'

The commission has not prohibited or created any bar for technological advancements and productivity. Instead, then it has emphasized on finding a right balance between environment and development. To maintain Harmony between environment and development was focussed. Technological advancement leads to better growth and enhancement of proficiency and productivity in the country. but at the same point of time, it will open the way for toxic waste generation. (Chowdury etal.,2024)

Montreal Protocol on Ozone Depleting Substances (1989)

In the year 1987 and International Agreement was made to reduce the import of Ozone Depleting substances. The major aim of this agreement is to protect the ozone layer of the Earth. There is a presence of harmful chemicals found in Refrigerators and Air conditioners. It consists of CFCS (chlorofluorocarbons) and HFCS (Hydro chlorofluorocarbons) and ODS (Ozone depleting Substances). The Montreal Protocol had successfully aimed in eliminating presence of greenhouse gases that are hampering the environment. ODS consists of 100 humanly produced chemicals. The irony is that when these chemicals get released in the environment deplete the ozone Layer.

Montreal protocol states about setting out different timetables for Developed and Developing Nations in relation to their Production and Consumption and in phasing out ODS properly.

London Guidelines for the Exchange of Information on Chemicals in International Trade, 1989

These guidelines were amended in 1989; It was formulated to increase chemical safety through bilateral talks between the nations. It has majorly focused in expanding the information in all areas i.e., scientific, technical economical and legal area. It has also laid emphasis on alternative methods and ban the usage of harsh chemicals. Hence, it has laid importance to methods of training which will eliminate the risk of accidents and there will be safe use of chemicals. Developed nations are prone to severe threats though these guidelines were still not formulated to address the issues of effluent nations but instead of that it provides effective framework that will provide exchange of information between the nations. (Arora etal.,2024)

E-Waste Under Basel Convention

The Basel convention on Tran's boundary movement of hazardous waste was signed by 173 countries on 22 March 1989 and entered into force on 5 May 1992. Annexure VIII is an annexure which covers another category of hazardous waste which is known as E-waste. Two step requirement process controls the hazardous process of E-waste under Basel convention. In the year February, 1998, fourth meeting of Conference of parties was held where the parties have emphasized upon the emerging problems related to E-waste. In the same conference List, A and B were included. List A and List B were signified as Annexure VIII and Annexure IX and it was rigorously prepared by technical working group. The convention was majorly aimed at preventing dumping of hazardous waste from developed to developing country which was the biggest challenge to overcome.

1. Scope of Basel convention

Basel convention mainly aims to control the movement of hazardous waste in an effective manner by ensuring that there must be collective effort and collaboration at the international front. Article 1 of the convention defines about the scope of waste. Many countries like India and Africa lacks technology to curb down E-waste.

2. Illegal shipping and Basel convention

Developed countries generate abundant amount of E-waste due to technological up gradation and advancement in the area of EEE

International Labor Organization Convention on Chemicals (1990)

ILO (no .170) The Labors use various sorts of chemicals at workforce. ILO focuses upon providing information about chemicals to the workers they had been using so far. The Employers and Suppliers are entrusted with the responsibility to provide accessibility of Information, training, and Assistance to these workers. It also lays emphasis that there must be proper coordination developed between workers and Employers. It has also laid down standards for classification, marking and labeling. Further, it also provides various measures to the employers such as First Aid, Medical facilities etc. (Recommendation no177)

Organization for Economic Cooperation and Development (1992): The developed Member Nations have entered into Council Decision waste Agreement that has focused and aimed to reduce the Trans boundary movement of Hazardous wastes. A comprehensive framework was established for OECD member states, to monitor and restrict the movement of wastes in OECD area.(Kaushik etal.,2024)

The Earth Summit, 1992

United Nations conference on environment and development also known as Earth summit was a global international environmental conference between 179 nations which was witnessed by Leaders, Politicians, Diplomats, and scientists. The summit had focused on achieving sustainable development by the people living at all levels. The summit is known for various achievements and documents which are produced

1. Rio Declaration
2. United Nations framework convention on climate change
3. Convention on biological diversity
4. The declaration on the principles of forest management
5. Agenda 21

(1) Rio Declaration

World commission on environment and development which is known as Brundtland commission was set up in 1983 by United Nations General Assembly. It was named on the name of its chairperson Go Harlem Brundtland who was the former Prime minister of Norway. The report was published in 1987 known worldwide as "Our common future" This report had emphasized to strike a perfect balance between economy and environment so in any case economic development would not pose a serious threat or risk to future generations. Hence, future generations would not be able to reap the fruits of earth.

Stockholm conference on environment and development had witnessed its twentieth anniversary which was celebrated by 178 nations including non- governmental organizations and other state parties met at Rio de Janeiro to discuss environmental problems which was creating lot of issues at international level. The conference again majorly emphasized on striking a balance between economic development and environmental concerns. (Singh etal.,2024)

United Nations conference on environment and development was brief enough to gather worldwide attention after three years which had inundated humans to live responsibly and cautiously by having best intentions towards environment.

Principle 2 of Rio declaration states that the activities which are undertaken within the jurisdiction of state does not hamper the environmental concerns of other states

Principle 15 is about Precautionary principle which will be hugely accepted by states to prevent environment

Principle 16 states that polluters should bear the cost of pollution without disturbing the process of international trade and investment

Principle 17 mandates Environment Impact Assessment (EIA) which had worked at the National level and had engaged in activities that will significantly reduce the impact of environmental hazards

It was October 17-21, 2011 when the nations gathered to witness the tenth Conference of Parties (COP) to the Basel convention which was aimed at reducing the movement of hazardous substances between the nations which had created adverse impact on the environment. COP,2011 was based on the very interesting Theme of "Prevention, Minimization and Recovery of waste" This conference was taken very substantially as it was believed that it will give significant contribution to United Nations conference on sustainable development (Singh etal.,2024)

Rio 20 is also known by the other name i.e., United Nations conference sustainable development which was held in 2012. The conference had three objectives which it seeks to achieve – There will some emerging and challenging situations in the relation to the environment which will be handled rigorously There are two themes which was agreed by member states i.e., Green Economy and sustainable development will go hand in hand by striking out adequate balance between the two and Poverty eradication and institutional framework on sustainable development.

1. United Nations Framework Convention on Climate Change

The UNFCCC aimed at various principles for reducing green house gas concentrations in the environment. It has also focused on reducing anthropogenic activities in the environment. Kyoto Protocol was adopted in December 1997, to increase the significance of UNFCCC. UNFCCC was adopted in the year 1992. The major aim was to reduce Hazardous activities that are creating interference with the climate system.

2. Convention on Biological Diversity

The United Nations Convention on Biological Diversity is one of among three international agreements based on the environment. This agreement is one among the agreement that has been envisaged from Rio Earth Summit held in the year 1992. It has focused on three objectives: Conservation of Biodiversity, Sustainable use of its components and Fair and Equitable sharing of benefits arising from the use of genetic resources.

3. The Principles of Forest Management

The Principles of Forest Management has focused that Nations have the right to seek benefit from their own Forest resources. It has also emphasized that it must be within the ambit of Forest Protection, Conservation and Management. These principles are extremely essential for sustainable development economy and crucial for maintaining

balance in all forms of Life. It emphasized that Nations must focus on 'Greening of the World' by planting more and more trees. and Harmful substances must be eliminated from the environment that is causing threat to Forests. (Dutta etal.,2024)

Agenda 21

Agenda 21 is a detailed comprehensive and non-legally binding document which has emphasized on population, consumption, and technology. It has focussed on the importance of sustainable development by balancing out consumption patterns in various parts of the world. Agenda 21 has various programmes and policies which had focussed on the importance of natural resources to build earths life supporting capacity.

This none legally binding document gave equal importance to both regulatory mechanisms and market mechanisms in a hope that it will reduce hazardous waste at all levels.

According to Article 21.10 of Agenda 21, Governments should initiate programmes to achieve sustained minimisation of waste generation and according to Article 21.9(b), by the year 2000, industrialised nations or developed nations must aim to prepare programmes that will reduce the production of wastes. Agenda 21 had focussed on waste recycling and reusing standards which will eliminate the risk of hazardous waste in environment. Efficient and sound management of e-waste will reduce the risk of threat and it will ensure to provide awareness and education to all age groups of society. (Ivkovic etal.,2024.)

Article 20.6 states that women and children are more prone to hazardous waste due to their biological condition due to which there must be conduction of research activities to find out the impact of hazardous waste on women and children

United Nations Framework Convention on Climate Change (1994): Toxic to Green initiative was taken up to create resilience among E-waste pickers in India. The E-waste pickers are trained and provided with assistance to handle and dispose of e-waste soundly. Further In addition, it also aims to create green jobs for E- waste pickers. And it also lends protection to E-waste pickers from the exposure of toxins. (Shrivastava etal.,2024.)

Conference on Chemicals Management (1995): ICCM has been working efficiently for the sound management of chemicals. It has created a mechanism to ensure safety of chemicals. However, to implement Strategic Approach to chemicals Management, a platform was provided to review and discuss the progress since its adoption in the year 2006. Even, world Summit on sustainable development, 2002 had the goal of sound and efficient management of chemicals.

Waigani Convention (1995): The convention was signed to put restriction on the export of radioactive wastes or Hazardous wastes to the countries that are members of Pacific Island forum. There has been a restriction laid on the export of Hazardous wastes or radioactive wastes.

Rotterdam Convention, 1998

Rotterdam convention had 163 parties and 72 signatories. India became the member of the convention on 24th may 2005. The theme of this convention was "Global Agreements for a healthy Planet: Sound Management of chemicals and waste. The convention has set out various objectives to achieve - like to have environmentally sound use of hazardous chemicals, Parties to this convention focus on putting up efforts in trading hazardous chemicals at the international level. This convention had listed out two very crucial mechanisms Prior Informed consent (PIC) and Information Exchange

Prior Informed consent is one of the main procedures to convention that lays down the procedure of importing chemicals in Annexure III. In Annexure III member parties have several choices to allow or disallow import of chemicals

INFORMATION EXCHANGE

In this mechanism secretariat will be informed about taking a regulatory action regarding banning of chemicals.

Millennium Development Goals (MDG) 2000

United Millennium development goals have agreed and aim to achieve 8 goals by the year 2015. In the year 2015 UNMDGS have achieved its goal of safe drinking water. These goals are to be enforced both by developing and developed nations. Its 7th Goal sets target to achieve environmental sustainability Since 1990 there has been constant increment in percentage of targets. In 1990 only 76% of population was able to have accessibility of safe drinking water However, in 2012 the percentage got increased by 14% and it has achieved target of 90%. Finally In the year 2015 United Nations Millennium Development goal has achieved the goal of safe drinking water. (Naushin etal.,2024.)

Millennium Development goals have not given sufficient emphasis on Environment issues. Environmental concerns were not paid any heed but, these issues required to address with an utmost concern and sensitivity. The concept of E-waste management is not even discussed. Most of the sustainable development goals revolve around combating diseases, Improving Maternity health, Eliminating Poverty and Hunger.

The Aarhus Convention (2001): The Convention was signed on 25th June 1998. This convention consists of 8 protocols that were established to eliminate Air Quality issues of the European Union. The Aarhus convention was based on three pillars i.e., Availability of Information, Participation of masses and Access to Justice.

The Johannesburg Declaration on sustainable development, 2002: It was in Johannesburg, 2002 when world summit on sustainable development was held to evaluate the progress of Rio Conference. India was an active participant of this declaration in 2002. The declaration has set goals to achieve in the areas of protection of biodiversity, sanitation, food security and access to health care facilities etc. There was a very dismaying moment for declaration when it failed to achieve goal on renewable energy. There was a coordinated approach taken by parties of the summit who worked together in ensuring financial assistance for creating development. Issues and concerns related to e-waste management were indirectly addressed by summit. However, due to lack of proper e-waste management when waste gets discharged in ground water and oceans, it thoroughly contaminates it.

Waste Electrical and Electronic Equipment (WEE) Directive in European Union

It was issued in 2003 as EU directive2002/96/EC which had made an intention to reduce electronic waste by encouraging recycling and reusing electronic products. It was observed that approximately one third of electronic waste is being recycled. This Directive was majorly focused upon the system where consumers were supposed to send the E-waste back and there was no amount charged for the same. Hence, it worked upon collection schemes and these collection schemes had thereby worked to increase the process of E-waste recycling and to ensure that Electrical and Electronic Equipment can be used again and again. (Naushin etal.,2024)

The Durban Declaration, 2008: The Durban Declaration on E-waste Management was held in Durban South Africa in the year 2008. The Durban Declaration and its signatories emphasized on the significance of extending the life of Electrical and Electronic Equipment (EEE). It has laid emphasis on the impact of WEEE and its disposal on Flora and Fauna. It has also highlighted the significance of Non-governmental Organizations, Local, Regional and Global partnerships that they address the issues of E-waste concerns.

In addition, there was a lesson taken from the experience of various countries such as Kenya, Morocco, and Uganda. There was a necessity generated to improvise the system of E-waste channelization in the continent of Africa. There were various highlights of this declaration: It is onus on the countries to monitor and scrutinize the prevalent E- waste management in their countries and put up some efforts to improvise E-waste channelization. Hence, there must be some efforts made in the direction of enforcement and compliance of E-waste management.

Libreville Declaration, Africa (2008): Libreville Declaration was the result of First Inter-Ministerial Conference held in Libreville Gabon in the year 2008 conference was based on Health and Environment. The landmark meeting was conducted by World Health organization, United Nations Environment Programme (UNEP), along with in association of Government of Gabon. The conference was witnessed by 52 Ministers of Health and Environment of different African Nations. It has been observed that there is further more research required to discover the threats or risks associated with the environment. It was also discovered that there can be vulnerabilities developed due to the risks associated regarding E-waste Management, for an instance, Like Poor Health and Lack of basic sanitation etc.

Hongkong International Convention for the safe and environmentally sound Recycling of ships (2009): The convention was adopted at International Maritime Organization (IMO) at Hongkong. The convention focused on replacing Harmful substances with non-Harmful substances that can be used for the purpose of refurbishment and re improvisation of ships. Further In addition, the convention had also tried eliminating the threats associated with ship recycling and occupational hazards.

Minamata Convention on Mercury (2013): This is a global treaty signed at Kumamoto Japan on 10 October 2013. The convention aims to reduce and eliminate the threatening effects of Mercury on Human Health and Environment. There have been various anthropogenic activities conducted in the environment, due to release of Mercury in the Environment that can cause adverse impact to the natural resources. There has been myriad of products where mercury has been found. The convention had laid restriction on those products where mercury has been found. It creates restriction on import and export by the year 2020. The restriction shall not be operational for those countries that had requested for relaxation for the initial 5-year period.

Waste Electrical and Electronic Equipment Regulations 2018 amendments (Open Scope): These regulations were implemented by all the member states of the European Union. It was in August 2005 when WEE came into force. It was in July 2007 since it got implemented in UK. The intrinsic goal of WEE is to eliminate and reduce the effects caused in the European union by EEE. The cost of recycling /handling of EEE are to be borne by the Producers. There was another version of EU directive that came into force in 2012, and later there were some amendments introduced that came into effect from 1st January 2019 onwards. WEE Regulations has acknowledged the amendment 'Open Scope 'were all EEE comes under the ambit of these regulations. There is total 14 categories of EEE that has covered Luminaires, Gas Discharge and LED Lamps. (Covered replaceable LED modules)

CONCLUSION

E-waste generation has been rapidly increased since past few years due to wide expansion and growth in Information and communication Technology sector. The statement and quantity generated in the past few years has been mentioned in various reports. The Researcher has highlighted and focused upon various theories that are based upon the issues of E-waste generation in developed and developing Nations. i.e., Race to bottom theory, Pollution Haven, and Distancing Theory etc. These theories helped in eliminating the causes of E-waste generation.

The significant contribution of United Nations and its entities are also included and their impact on E-waste has been analysed. E-waste and its Legislations across various countries and laws prevalent during current times have been critically analysed. It is extremely imperative for the countries to take steps to build significant Legislations If deterrent theory is adopted by the countries in environmental concerns, and then the individuals must be dispirited from committing crimes in future.

'Right to Repair' is a newly added concept due to lack of transparency on the part of Manufacturer. Its Regulations in USA and European Union have been researched. The major aim of 'Right to Repair' is to lessen the burden on the consumers to purchase new products. The consumers and Purchasers need to ensure they are repairing and re improvising the product on its own.

REFERENCES

Ali, Z., Jamil, Y., Anwar, H., & Sarfraz, R. A. (2024). Classification of e-waste using machine learning-assisted laser-induced breakdown spectroscopy. Waste Management & Research.

Arora, M. K., Lal, S., Singh, B., & Olyaee, M. (2024). Through an Energy Lens: Examining SDGs From a Policy Perspective. In Riswandi, B., Singh, B., Kaunert, C., & Vig, K. (Eds.), *AI Applications for Clean Energy and Sustainability* (pp. 320–343). IGI Global. DOI: 10.4018/979-8-3693-6567-0.ch016

Ashraf, W. M., Jadhao, P. R., Panda, R., Pant, K. K., & Dua, V. (2024). Towards circular economy of wasted printed circuit boards of mobile phones fuelled by machine learning and robust mathematical optimization framework. *Resources, Conservation & Recycling Advances*, 23, 200226. DOI: 10.1016/j.rcradv.2024.200226

Ashraf, W. M., Jadhao, P. R., Panda, R., Pant, K. K., & Dua, V. (2024). Towards circular economy of wasted printed circuit boards of mobile phones fuelled by machine learning and robust mathematical optimization framework. *Resources, Conservation & Recycling Advances*, 23, 200226. DOI: 10.1016/j.rcradv.2024.200226

Belsare, K., Singh, M., Gandam, A., Malik, P. K., Agarwal, R., & Gehlot, A. (2024). An integrated approach of IoT and WSN using wavelet transform and machine learning for the solid waste image classification in smart cities. *Transactions on Emerging Telecommunications Technologies*, 35(4), e4857. DOI: 10.1002/ett.4857

Debnath, B., Chowdhury, R. R., Acharya, C., & Guha, A. (2024). Sustainability Aspects of E-waste Supply Chain Network Role of New and Emerging Technologies. Technological Advancement in E-waste Management, 63-79.

Dutta, P. K., Singh, B., Towfeek, A.-S. K., Adamopoulou, J. P., Bardavouras, A. N., Bamwerinde, W., Turyasingura, B., & Ayiga, N. (2024). IoT Revolutionizes Humidity Measurement and Management in Smart Cities to Enhance Health and Wellness. *Mesopotamian Journal of Artificial Intelligence in Healthcare*, 2024, 110–117. DOI: 10.58496/MJAIH/2024/013

Etensa, T., Alemu, T., & Yayo, M. (2025). Rethinking the measurements and predictors of environmental degradation in Ethiopia: Predicting long-term impacts using a kernel-based machine learning approach. *Environmental and Sustainability Indicators*, 25, 100554. DOI: 10.1016/j.indic.2024.100554

Farshadfar, Z., Mucha, T., & Tanskanen, K. (2024). Leveraging Machine Learning for Advancing Circular Supply Chains: A Systematic Literature Review. *Logistics (Basel)*, 8(4), 108. DOI: 10.3390/logistics8040108

Gnanasekaran, C., Govindaraj, M., Jayavelu, S., & Chandrasekar, V. (2024, November). IoT and strategic management in action: Pilot testing for efficient e-waste collection in rural Indian households. In AIP Conference Proceedings (Vol. 3192, No. 1). AIP Publishing.

Ivković, R., Petrović, M., Spalević, P., & Milivojević, Z. (2024). Classification of E-Waste Types Using Machine Learning and Digital Image Processing. *Przeglad Elektrotechniczny*, 2024(9).

Joshi, P., Bagul, P., & Deshpande, A. (2024, October). Towards Sustainable E-Waste Management: A Computer Vision and Machine Learning Approach for Identification and Valuation. In *2024 International Conference on Intelligent Systems and Advanced Applications (ICISAA)* (pp. 1-6). IEEE. DOI: 10.1109/ICISAA62385.2024.10828717

Kaushik, T. K., Singh, A., Raghav, A., & Singh, B. (2024). Potential of Artificial Intelligence in the Agricultural Industry: A Comprehensive Review. In B. Riswandi, B. Singh, C. Kaunert, & K. Vig (Eds.), AI Applications for Clean Energy and Sustainability (pp. 344-365). IGI Global. DOI: 10.4018/979-8-3693-6567-0.ch017

Lal, S., Arora, M. K., Riswandi, B. A., Singh, N., & Singh, B. (2024). Melding Artificial Intelligence and Remote Sensing Data for Real-Time Monitoring of Marine Biodiversity. In Riswandi, B., Singh, B., Kaunert, C., & Vig, K. (Eds.), *AI Applications for Clean Energy and Sustainability* (pp. 299–319). IGI Global. DOI: 10.4018/979-8-3693-6567-0.ch015

Naushin, M., Saraswat, A., & Abhishek, K. (2022). Deep learning approach to deal with e-waste. In *Advanced Machine Intelligence and Signal Processing* (pp. 123–136). Springer Nature Singapore. DOI: 10.1007/978-981-19-0840-8_9

Oise, G., & Konyeha, S. (2024). E-Waste Management through Deep Learning: A Sequential Neural Network Approach. *FUDMA JOURNAL OF SCIENCES*, 8(4), 17–24. DOI: 10.33003/fjs-2024-0804-2579

Oise, G. P., & Konyeha, S. (2024). Harnessing Deep Learning for Sustainable E-Waste Management and Environmental Health Protection.

Oise, G. P., & Susan, K. (2024). Deep Learning for Effective Electronic Waste Management and Environmental Health.

Raghav, A., Singh, B., & Jermsittiparsert, K. (2024). Role of Artificial Intelligence and Machine Learning in Clean and Blue Energy for Enhancing Efficiency and Sustainability. In Riswandi, B., Singh, B., Kaunert, C., & Vig, K. (Eds.), *AI Applications for Clean Energy and Sustainability* (pp. 366–386). IGI Global. DOI: 10.4018/979-8-3693-6567-0.ch018

Ramya, P., Ramya, V., & Babu Rao, M. (2024). Optimized deep learning-based e-waste management in IoT application via energy-aware routing. *Cybernetics and Systems*, 55(8), 2041–2070. DOI: 10.1080/01969722.2023.2175119

Singh, B. (2024). Cherish Data Privacy and Human Rights in the Digital Age: Harmonizing Innovation and Individual Autonomy. In Pucelj, M., & Bohinc, R. (Eds.), *Balancing Human Rights, Social Responsibility, and Digital Ethics* (pp. 199–226). IGI Global. DOI: 10.4018/979-8-3693-3334-1.ch007

Singh, B., & Kaunert, C. (2024). Human-Machine Nexus for Digital Rebound Fostering Futuristic Energy-Efficiency: Ecological Footprint of Technology for Smart-Sustainable Urban Mobility. In González-Lezcano, R., & Sansaniwal, S. (Eds.), *Design Strategies for Efficient and Sustainable Building Facilities* (pp. 222–244). IGI Global. DOI: 10.4018/979-8-3693-3200-9.ch010

Singh, B., & Kaunert, C. (2024). Hidden Gems Breakthrough Dynamic Landscape of Adaptive Learning Technologies for Higher Education: Bridging the Gap Between Theoretical and Practical Knowledge Projecting Student Learning Outcomes. In Minh Tung, T. (Ed.), *Adaptive Learning Technologies for Higher Education* (pp. 222–247). IGI Global. DOI: 10.4018/979-8-3693-3641-0.ch010

Singh, B., & Kaunert, C. (2024). Embryonic Machine-Deep Learning, Smart Healthcare and Privacy Deliberations in Hospital Industry: Lensing Confidentiality of Patient's Information and Personal Data in Legal-Ethical Landscapes Projecting Futuristic Dimensions. In Kumar, P., Singh, P., Diwakar, M., & Garg, D. (Eds.), *Healthcare Industry Assessment: Analyzing Risks, Security, and Reliability. Engineering Cyber-Physical Systems and Critical Infrastructures* (Vol. 11). Springer. DOI: 10.1007/978-3-031-65434-3_7

Singh, B., & Kaunert, C. (2024). AI-Driven Strategies for Customer Engagement, Market Segmentation, and Resource Optimization: Projecting End User Satisfaction and Futuristic Growth of Business. In Nadda, V., Tyagi, P., Singh, A., & Singh, V. (Eds.), *AI Innovations in Service and Tourism Marketing* (pp. 104–128). IGI Global. DOI: 10.4018/979-8-3693-7909-7.ch006

Srivastava, S. K., & Shrivastava, R. K. (2023). Advanced Approach towards Zero Waste: Modeling of Copper Recovery from e-Waste by Using Machine Learning Technique.

Yang, L., Zhang, T., Gao, Y., Li, D., Cui, R., Gu, C., Wang, L., & Sun, H. (2024). Quantitative identification of the co-exposure effects of e-waste pollutants on human oxidative stress by explainable machine learning. *Journal of Hazardous Materials*, 466, 133560. DOI: 10.1016/j.jhazmat.2024.133560 PMID: 38246054

Chapter 7

Revolutionizing Rural Development: Mobile-GIS and Cognitive Intelligence Systems - A Case Study of Chhattisgarh

Anand Tamrakar
https://orcid.org/0009-0001-9213-7404
MATS University, India

K. P. Yadav
MATS University, India

ABSTRACT

The integration of artificial intelligence (AI) with Mobile-GIS is revolutionizing rural development in Chhattisgarh by providing innovative solutions across agriculture, healthcare, education, and resource management sectors. Predictive analytics for smarter agriculture, environmental sustainability, and better decision-making are just some of the ways in which AI-powered cognitive systems are making a difference. The companies have been offering technology such as Mobile-GIS, accurate mapping, real-time monitoring to deploy effective resource allocation and disaster management. These technologies, used in unison, can give rural communities data-driven insights and tools to enhance livelihoods and quality of life. Early implementations in Chhattisgarh show promise, with greater agricultural productivity and public service delivery streamlined if data is shared freely. Combining AI and GIS can help Chhattisgarh evolve into a model of future-ready, inclusive, efficient, sustainable rural development.

DOI: 10.4018/979-8-3693-9410-6.ch007

1. INTRODUCTION

Affairs Evaluation Artificial Intelligence Artificial Intelligence In Rural Development GIS GIS In (in) India (in) India The rural development landscape in India is going through a paradigm shift owing to the fast-paced advancements within the area of Artificial Intelligence (AI) and Geographic Information Systems (GIS). With an abundance of natural resources and predominantly rural demography, Chhattisgarh one of the states in India experiences opportunities and challenges in the implementation of these technologies for Sustainable Development which are specific to the state's socio-economical and geographical context. In addition, cognitive intelligence elements of AI and Mobile-GIS technologies are helping the state to solve systemic issues of agriculture, healthcare, education and resource management, which is essential for state's socio-economic development. From its ability to make real-time analysis accessible in even the most far-flung corners of the globe to its potential for delivering tailored applications that mirror surgical precision, these technologies will undoubtedly become a game-changer in rural development and help bridge the rural-urban divide for holistic living (Griffin et al., 2024).

But rural Chhattisgarh, with its vast agricultural base and forested regions, is a microcosm of the problems that confront rural India. The scarcity of infrastructure, low levels of literacy, inadequate availability of medical services and lack of efficient management of natural resources continue to inhibit growth. The heavy in-farm dominated population almost entirely depends on traditional harvest methods, the outcome of which is often likely to be bent by erratic weather and price distortions in the market. Additionally, environmental capacity, with regard to access to vital services, education, health care, and modern technology is hindered due to the isolation of many villages. It emphasizes challenges such as the crucial need for transformative solutions addressing the symptoms and root drivers of underdevelopment.

Sticking to Artificial Intelligence comes with a unique potential that is precisely capable of addressing these challenges. For instance, AI powered cognitive systems are able to digest large sets of data process them using analytics and derive insights to make better decisions. For instance, in agriculture, predictive analytics can be used to forecast crop yields, facilitate optimal irrigation times or provide alerts for pest conflicts in advance. These types of applications fuel improvements in agricultural productivity, and thus food security and sustainable resource utilization. AI-powered applications have, thus, enabled better allocation of resources in health care and education, making it feasible to deliver vital services to underserved rural communities. Supply chain efficiency and AI, through automatic data collection, real-time data processing, smart routing, and predictive analytics, has greatly improved supply chain operations across the globe.

AI brings computational power, but Mobile-GIS technology embeds spatial intelligence at the core of rural development. GIS applications provide actionable insights into resource distribution, land use planning and disaster management by mapping and visualising data over geographies. In offline rural situations where quick choices are crucial, Mobile-GIS helps with real-time monitoring and data collection. These technologies have huge significance in Chhattisgarh in managing natural resources like water, forests and minerals. Besides this, the GIS tools are useful to map up the more underdeveloped areas, check for lost infrastructure in the area and find a way around those so that these barriers are exceeded.

AI and Mobile-GIS when integrated in one marketplace creates a synergised ecosystem contributing to the welfare of mankind through solving multidimensional problems on ground in an all encompassing way. In Chhattisgarh, this integrated approach can provide huge socio-economic benefits. As an example, AI algorithms can analyze GIS generated spatial data to identify high-risk areas that require immediate intervention (improvement of infrastructure or disaster relief). And, in a like manner, Mobile-GIS can augment AI-based models, providing real-time geospatial analytics to add depth and context to results. These technologies put together can fine-tune various processes such as rural electrification, water management, government scheme success and optimal utilisation of resources.

Even though AI and GIS technologies have immense potential, the adoption of such technologies in the rural belt of Chhattisgarh presents significant challenges. Most villages have very poor technological infrastructure, internet and power supply is only constant. Additionally, expertise is required to deploy cutting-edge technology, and it is in short supply in rural areas. In addition, the adoption of technology is more complicated and delayed due to various cultural and social dynamics (Shakhislam et al., 2024), leading to a continual resistance against technological change. The more ethical the data collection process will be, the more successful one can be in establishing trust with local communities, since privacy and security issues are key factors there.

A multi-stakeholder perspective is critical to overcoming these barriers. A harmonious ecosystem led by collaboration among government agencies, private sector interests, and civil organizations that will allow for effective deployment of AI and GIS in the field. There needs to be greater policy frameworks to give the infrastructure and incentives for the technology to be adopted. In addition, education and training centers can provide local people with the abilities necessary to maximally use these techniques. And also, such campaigns can help demystify AI and GIS, encouraging communities embrace their transformative potential.

Well, other than that, the socio-economic impact analysis of the adoption of AI-powered cognitive systems and Mobile-GIS in rural Chhattisgarh is far-reaching. On the other hand, new job scratches in fields like data review, system sustainment, and

GIS socio-economic stems may arise due to these technologies. In this way, they can enable rural people to participate actively in development activities by promoting skills and knowledge transfer. Accessibility to better healthcare, education, and sustainable resources can improve the living standards of most of the rural community thus ensuring the urban-rural divide is bridged. The integration of AI and GIS in environmental monitoring also plays a crucial role in disaster preparedness and response, reducing the socioeconomic impact of natural disasters and enhancing the resilience of at-risk individuals.

Government initiatives are pivotal in itself in scaling these innovations. The Digital India and AI for All programs of India provide the basic building blocks to drive technology adoption in the rural ecosystem. At the state level, in Chhattisgarh, more can be done in conjunction with the national level, to address the issues specific to the region. In doing so, encouraging private-sector investments in AI and GIS solutions and partnering with academic and research institutions can foster the modeling, development and deployment of tailored solutions. These pilot projects and case studies in few selected villages would be examples to scale it up in other villages of the state (Aydinoglu et al, 2024).

So, as Chhattisgarh takes the first step toward extracting the potential of AI and GIS for rural development, ethical considerations must be placed at the center. Data privacy, AI algorithm bias, and community participation are essential to building relationships and the prolonged success of efforts. We need to develop a legal and regulatory framework that adapts to the challenges of these technologies while maintaining a balance between innovation and accountability. These technologies have the potential to transform rural development by addressing key challenges and unlocking latent opportunities for sustainable economic growth and enhanced quality of life. But achieving this vision will depend on all stakeholders aligned behind policies and community engagement and with a commitment to innovation. With Chhattisgarh utilizing cutting-edge tools for rural development, the move can serve well as a model for the world to undertake equitable rural development with the help of technology.

2. CURRENT CHALLENGES IN RURAL DEVELOPMENT IN CHHATTISGARH: SOCIO-ECONOMIC LANDSCAPE

Chhattisgarh is the name of one of India's richest resources but its poorest state. More than half of the population lives in rural areas and are mostly engaged in agriculture and allied activities. Yet rural poverty is aggravated by limited access to modern agricultural practices, volatile markets and inconsistent income. Unfortunately, the majority tribal population of the state faces social exclusion, dearth of education and healthcare accessibility leading to further aggravated inequality. Lack of diversification in employment opportunities encourages many to either work for subsistence or migrate to urban areas (Xie et al., 2024). Out of which primarily women are illiterate which obstructs the women empowerment, economy and struggle towards breaking the cycle of extremes of poverty.

2.1 Infrastructure Deficiencies

The state of Chhattisgarh is still light years behind in rural development due to egregiously inadequate infrastructure. We still lack quality transportation networks as many villages are cut off from the rest of the world, and access to markets, healthcare, and even education remains limited. Delayed delivery of agricultural produce because of inadequate road connectivity and poor organised structure for marketing leads to losses in the first place and reduces income, discouraging farmers from increasing productivity. Moreover, there are disparities in electricity and internet penetration in the rural countryside, making technological adoption and modern educational tools less available. Healthcare infrastructure is another vital aspect, as health centers in rural areas are often under-staffed and lack required equipment to serve the needs of the population. This is compounded by the lack of clean drinking water and sanitation facilities leading to adverse health effects. This is an interminable cycle in which inadequate educational institutions and infrastructure persistently deprive rural youth and children of opportunity for self-improvement. Addressing these infrastructural issues is essential for bridging the rural-urban divide in Chhattisgarh (Kaldeli et al., 2021).

2.2 Resource Management Issues

The region's abundant natural wealth, including its forests and minerals, has been overshadowed by mismanagement and violence. The overexploitation of these resources and the ineffectiveness of management practices has led to environmental degradation that threatens the livelihoods of the individuals who rely on the land and the forests. Mining activities are essentially unregulated and have become both a boon and bane for the environment as they suspend ecosystems and displace local communities, leaving the soil and water contaminated. Additionally, irregular patterns of rainfall, low-levels of infrastructure, and lack of irrigation systems have made management of water resources an indefinite challenge for the agriculture to be harvested. The problem of wastage and sustainability is further aggravated by the massive wastage of resources owing to the traditional and inefficient means of resource utilization, particularly in rural areas. Traditional rights of forest-dwelling communities that are intricately linked to the ecosystem are often ignored, leading to social economic displacement. Resource management is a complex issue that, if handled well, could have positive impacts in, and beyond, Chhattisgarh (Rahimi et al., 2021).

3. ROLE OF AI IN REVOLUTIONIZING RURAL DEVELOPMENT

Artificial Intelligence (AI), specifically cognitive intelligence systems, holds significant promise in tackling challenges faced in rural development. They emulate human thought and problem-solving skills, enabling them to evaluate extremely complicated datasets and produce actionable insights. In Chhattisgarh, these systems can help bring efficiencies across a number of development processes. Examples show that cognitive intelligence can be used to design targeted government schemes and identify the needs of various communities. Using socio-economic data, these systems can highlight the interventions that would have the most impact, directing scarce resources to those most in need. In education, they can localise content for language and cultural contexts, improving accessibility and engagement. Integrating cognitive systems into rural development frameworks would enable whatever policy Chhattisgarh may formulate would only be relevant in a data-oriented and integrated ecosystem.

3.1 Agriculture Predictive Analytics

Agriculture, the backbone of rural Chhattisgarh, can stand to gain a lot by such measures. These algorithms study past records, meteorological data, soil conditions along with the market trends to enable farmers with tangible recommendations. Predictive models can project crop yields, allowing farmers to schedule harvests and allocate resources accordingly. They can also anticipate outbreaks of pests and diseases, enabling timely intervention and minimizing crop damage. Moreover, predictive analytics can refine irrigation schedules according to current weather information, saving water and improving efficiency. Such tools help famers with knowledge and transition from traditional practices by providing the farmers the tools to enable more efficient and sustainable farming practices. Building on the predictive capabilities of data analytics, similar approaches can empower farmers in Chhattisgarh, who are often challenged with uncertain rain patterns and market viability to make agriculture a much more assured and lucrative way of life, positively impacting food security and financial stability of communities alike (Force, 2022).

3.2 Improving the Decision-Making Process

Because of AI's ability to process and analyze massive amounts of data, it is an invaluable tool for enhancing decision-making in rural development. AI-driven platforms can analyse various dimensions of data surfaces, be it socio-economic, infrastructural, environmental, etc., to identify developmental paradigms for government agencies and policymakers. In Chhattisgarh, these systems can be used to prioritize distributing resources for constructing schools, hospitals, or water supply systems. AI tools in disaster management can also be used to predict natural calamities like floods or droughts and help devise mitigation strategies. You can also track implementation of the development programs through these systems, giving you live feedbacks and enabling the course corrections. AI has the potential to not only enhance decision-making processes that pave the way for effective rural development initiatives but also help in bringing the benefits to the underprivileged population of Chhattisgarh to reflect equity in rural development initiatives.

A path is opened towards the usage of AI powered cognitive intelligence and predictive analytics along with fact-based decision-making becoming a game changer and enabling rural development in Chhattisgarh. Despite ongoing obstacles in socio-economic dynamics, infrastructure, and resource management, these technologies have the potential to catalyze transformative change across all sectors. Utilizing AI, Chhattisgarh can establish an inclusive and sustainable development model that addresses its unique issues while improving the quality of life for its rural inhabitants. This transformation cannot occur in isolation; it requires collective action, inclusive policies, and an emphasis on innovation to guarantee that no citizen is marginalized in the pursuit of equality.

4. MOBILE-GIS SYSTEMS: A CATALYST FOR SPATIAL DEVELOPMENT

Mobile GIS applications facilitate precise mapping and efficient resource allocation and analysis by transforming the development of spatial environments. In underdeveloped rural environments like Chhattisgarh, where the inequality of resources determines the inequalities in development, GIS mapping has made a significant difference in addressing issues of inequality. GIS spatial data integration with demographic, economic, and environmental information enables targeted resource allocation to the most underserved regions through developmental interventions. GIS mapping, for instance, can help detect places with inadequate access to education, healthcare, and clean water, thus allowing policymakers to focus on developing infrastructure in these areas (Radosevic et al., 2023). Additionally, it aids in making agricultural resources more efficient by monitoring soil quality, water resources, land use, etc., which are the basis of sustainable farming. Furthermore, given that a large proportion of the population in Chhattisgarh still relies on agriculture, the use of GIS technologies and infrastructure for agriculture, in agriculture resources allocations will increase the menagerie of holistic rural development, by indicating for different institutional mechanism to enhance productivity and improve livelihoods.

Real-time monitoring capabilities have revolutionized the data collection and analysis process in modern mobile Geographic Information Systems (GIS) technology. Mobile GIS systems using GPS and IoT devices allow field workers to collect the data directly from the ground, saving time and energy needed for individual survey works. This is especially useful for Chhattisgarh because many villages in the state are geographically isolated and have no reliable communication networks. Updates can be provided in real-time by collecting data on critical parameters like crop health, water levels and disease outbreaks, allowing for timely strikes. Mobile GIS can also allow tracking of development projects progress, which can ensure transparency and accountability. The use of GIS-based tools can drastically improve the monitoring of the schemes execution like rural electrification or construction of roads to close this gap between planning and implementation of such complimentary schemes in a rural development initiatives (Xin & MacEachren, 2020).

Like Chhattisgarh many other places are eerily affected by floods, forest fires and droughts which JSON can change the game for the villagers in these areas. Mobile GIS is essential for disaster management for providing accurate spatial and non-spatial data for assessment, preparedness, and response. GIS mapping also helps in mapping high-risk zones as well as vulnerable populations that can be addressed through proactive disaster management techniques. Real-time Geographic Information System (GIS) data assist disaster rescue and relief coordination by determining safe routes, as well as locating affected populations and redistributing resources in an efficient manner. GIS tools are particularly useful when conducting damage assessments and planning for recovery in the aftermath of the disaster. For instance, after a flood, GIS can visualize the extent to which an area was inundated, assisting authorities to navigate rehabilitation and reconstruction of essential infrastructure. Mobile GIS will thus be integrated into disaster management framework will go a long way in building a resilient society minimizing the disaster impact on the rural class in Chhattisgarh. (Sailer, 2020).

5. INTEGRATING AI AND MOBILE-GIS IN RURAL DEVELOPMENT

Innovative case studies emerging from Chhattisgarh indicate that an amalgamation of Artificial Intelligence (AI) & Mobile-GIS systems may be adopted to extract the potential of data in rural development. AI based GIS platforms have been deployed for tracking forest resources (the state has one of the largest tribal populations in the country and depends heavily on forest based livelihoods). These platforms use satellite images and machine-learning algorithms to detect illegal logging and forest degradation to sustainably manage resources. As an example of this, AI-GIS systems have been used in agriculture to identify soil health and crop suitability patterns, resulting in increased crop yields and reduced environmental impact. In healthcare, for instance, mobile GIS with AI has allowed for the mapping of healthcare facilities and recording of diseases for on time action. The above practical case studies of leveraging AI and GIS together put light on the operational feasibility of tapping the potential of Ai-GIS integration to resolve the specific challenges of rural Chhattisgarh.

Integrating AI with Mobile-GIS augmenting both on their independent abilities is useful for rural development. Together, AI and GIS significantly improves the analysis capabilities of both technologies as AI automates the processing of data and generating predictive insights, and GIS offers the spatial context that is important for designing solutions and informed decisions. Combine to allow for greater targeted and timely planning, monitoring, and delivery of development projects. One such sector that benefits from integrated developments is in water resource management, where AI-GIS systems can be used to ascertain groundwater levels and allow for efficient irrigation scheduling, facilitating sustainable consumption. Integration helps in identifying underserved areas and better allocating resources in rural healthcare (Aydinoglu et al., 2024).

5.1 AI-Enabled GIS Workflows for Problem Solving

AI-GIS workflows provide a methodical problem-solving framework for rural development issues by integrating geographic information system (GIS) methodologies with artificial intelligence (AI) techniques. These workflows are typically a set of steps for acquiring, processing, analyzing, and making a decision with data and the integration of AI and GIS technology optimizes such processes. For example, data from satellite imagery and Internet of Things (IoT) sensors are aggregated and processed by AI algorithms for crop health identification and yield prediction, as well as information visualization on geographic information systems (GIS) for business intelligence in agriculture. Similarly in disaster management live data captured through drones and sensors are analyzed through AI models using these models to predict disaster scenarios while geographical data layer is overlaid using GIS for planning and activating a response. In Chhattisgarh, the natural representation of these workflows has been utilized to solve problems such as water scarcity, deforestation, rural connectivity etc., proving their versatility and effectiveness. These challenges can be addressed through some innovative, sustainable, and inclusive rural development solutions provided through implementation of AI-GIS workflows in Chhattisgarh (Zhao, 2022).

The integration of Mobile GIS systems with AI has been a catalyst for spatial development providing out-of-the-box solutions to the rural development scenario of the state. From resource allocation and real-time monitoring to disaster management and healthcare, such technologies have shown their transformative potential across sectors. Utilization of AI-GIS convergence can help Chhattisgarh make the best use of its resources, improve the making of informed decisions and be resilient to future adversities, thereby to facilitate sustainable rural development. But the successful implementation of these technologies will require collaboration, policy, and capacity development so that their benefits are shared in even the most distant corners of the state.

6. SOCIO-ECONOMIC IMPACTS OF AI AND GIS ADOPTION

The intersection of AI and GIS in the domain of rural development becomes more relevant in the context of Chhattisgarh, a state with socio-economic inequalities and developmental imbalance. These technologies not only do things like help manage resources and plan policies — when connected, they enable rural communities and provide jobs and improvements to everyday living. This would enable those in rural communities, who previously had no access to services and information, to learn. A GIS-based crop management system can also allow farmers to better understand potential soil conditions, the availability of water, and even weather patterns when deciding on planting, harvesting, and irrigation. Same goes for AI-powered platforms to analyze market trends in real-time for farmers, which can lead to better prices for their produce and lesser reliance on middlemen. AI-based e-learning portals and GIS mapping of schools are helping to bridge the gaps in accessibility in terms of education (Li et al, 2022) of students even in the remotest of villages. Use of GIS also helps local governments and community organizations to have a better insight into and track development where the development is taking place, who is being affected, such that development funds and resources are delivered and targeted to the populations who can realistically benefit from them.

6.1 Job Creation and Skill Development

Re-envisioning Rural Livelihoods with AI and GIS AI and GIS offer a range of new job opportunities and skill sets to work on and develop. From IT to agriculture to infrastructure development, jobs can also be created with the collection, analysis, and maintenance of systems powered by these technologies. This initiative not only helps bridge the digital divides that exist between urban and rural areas but also opens up pathways for careers in emerging technologies such as GIS mapping, AI analytics, and IoT device management. Moreover, an AI-empowered system can have the capability to recognize skill gaps present in communities, as well as suggest pertinent and customized training programs, ensuring that individuals are trained to participate in the new workforce. Therefore, the speculative and productive potential, represented through the innovative spectrum possibilities - as provided by AI and GIS - not only creates new jobs, but also revamps existing jobs through upskilling, thereby providing added economic resilience to rural populations through time (Yuan, 2024).

6.2 Improved Quality of Life

AI and GIS technology can introduce new paradigms in rural sectors which would enhance the quality of life by solving the critical issues about medical health care, basic education, infrastructure, etc. The synergy of GIS-based mapping of healthcare facilities with AI-driven diagnostic tools facilitates the efficient delivery of medical services, especially in the remote regions. If you take the case of telemedicine platforms, the start-ups can make precise diagnoses through AI's recommendations and even predict outbreaks for timely intervention. In academia, AI algorithms enable tailored learning experiences, personalized to the unique needs of each student and fostering better academic results. In addition, GIS mapping helps with infrastructure development by pinpointing the most suitable locations for developing roads, schools and water supply systems, ensuring that the projects are not only cost-effective but also impactful. These changes are transformative for rural communities and promote equity and sustainability for their residents and well-being.

6.3 Government Initiatives and Policy Framework

And the Indian Government has made steady and commendable progress in harnessing technology for rural development especially with the aid of Artificial Intelligence (AI) and Geographic Information Systems (GIS). These initiatives seek to bridge the developmental gap between urban and rural sectors, ensuring that growth is inclusive and sustainable. The Government, in acknowledgment of the transformational power of new technologies like AI and GIS, has also launched several initiatives and policies to support the adoption of AI and its integration into rural development programs (Arin et al., 2024).

The creation of a national foundationIn India, the Digital India program is an umbrella initiative for driving technology in the rural sector. Introduced in 2015, this initiative is designed to make India a digitally empowered society and knowledge economy. Digital India seeks to create a conducive environment for the adoption of technologies like AI and GIS by promoting digital infrastructure and better internet connectivity and digital literacy. Also, the National Artificial Intelligence Strategy focuses on the use of AI in three major sectors: agriculture, healthcare and education. The policy is supportive of AI in solving issues such as resource allocation, crop monitoring, and access to quality healthcare in rural regions.

Several GIS-based projects are also being undertaken by the Ministry of Rural Development in order to facilitate resource management and infrastructure development. For instance, the Mahatma Gandhi National Rural Employment Guarantee Act (MGNREGA) employs GIS mapping to track ongoing projects, ensure accountability, and optimize resource allocation. Another applications of GIS is the Pradhan Mantri Gram Sadak Yojana (PMGSY), which uses GIS to map the rural road networks to identify unconnected areas and helps for proper planning for connecting them. These efforts reflect the government's commitment to utilizing GIS as a tool to improve rural infrastructure and development outcomes.

Chhattisgarh has emerged as a State that overcame geographic and socio-economic challenges with the help of AI and GIS technologies to build its rural landscape. Policies at the state level are related to using GIS on schemes for agriculture, health and education (Li, 2023). For example, the Chhattisgarh Infotech Promotion Society (CHiPS) in India is instrumental in implementing e-governance projects and promoting the use of GIS for applications such as land management, asset and resource allocation, and disaster preparedness. The state will use technology to maximize developmental benefit and even out resource allocation.

Projects such as GIS-based agricultural management systems have also been implemented that enable farmers taking data-driven decisions. They offer information on soil health, rainfall, weather conditions, and access to water to ensure sustainable agricultural production. Chhattisgarh has also experimented with artificial intelligence-enabled health care solutions like telemedicine platforms to improve access to medical services in remote villages. These specific initiatives demonstrate the state's proactive approach towards embedding technology within rural development initiatives.

Although existing policies and initiatives form a solid ground, improvements are needed to increase the potential of AI with GIS in rural development. Second, a massive scaling up of investments in digital infrastructure to ensure that even the most remote areas are served with high-speed internet and reliable power supply. Second, there is a need for capacity-building programs to familiarize local communities and government officials with how to leverage AI and GIS technologies. Third, the private sector's expertise and resources can complement the public sector by being promoted through public-private partnerships (Xia, 2023).

The policy should also concern how challenges within data security and privacy are addressed, and how the rights of individuals are safeguarded. Another key step will be to establish ethical guidelines governing the use of AI and GIS, ensuring that these technologies are used in a way that promotes transparency, fairness, and accountability. An effective approach would be that the government and community need to fine-tune the policy framework and work on these critical areas to create an enabling environment for the adoption of the AI and GIS on its widespread basis in the rural development planning process. Government programs and policies have provided a solid basis for harnessing the power of AI and GIS in achieving rural development goals. These technologies hold the potential to revolutionize rural India as a hub for sustainable growth and creativity if only we can scale up existing initiatives and fill the gaps that still remain.

7. REGULATORY FRAMEWORK AND POLICIES PROMOTING AI AND GIS

The Indian government has acknowledged the potential of AI and GIS in transforming rural development and has implemented numerous policies and initiatives to foster the integration of these technologies. The Digital India campaign and National AI Strategy of India highlight the critical role played by technology in bridging developmental gaps and promoting inclusivity in the country. Projects such as those run by the Ministry of Rural Development also use GIS for resource mapping, land management, and disaster preparedness. For instance, in Chhattisgarh, state-specific policies place an emphasis on how to integrate GIS in rural development schemes, including the Mahatma Gandhi National Rural Employment Guarantee Act (MGNREGA) and the Pradhan Mantri Gram Sadak Yojana (PMGSY). These efforts reflect the government's initiative to deploy AI and GIS for sustainable development. Instead, they require a deluge of coalescing policies that work together to improve infrastructure deficits and digital literacy, overcome issues of data security and fund itself (Xia, 2022).

7.1 Policy Recommendations for Improving

For the full leveraging of AI and GIS technologies in rural development to be A reality, policies would require to be reorganised and new frameworks to be introduced. Firstly, the government should focus on making investments in digital infrastructure and ensuring that rural areas have access to high-speed internet and uninterrupted power supply. Second, capacity-building programs must be expanded to train local communities and government officials on the role of AI and GIS technologies. Third, policies should foster public-private partnerships, capitalizing on the capabilities and resources of the private sector to stimulate technology uptake and address challenges. Moreover, stricter data privacy and security measures should be put in place to ensure the trust of all parties involved and to protect sensitive data. These key enablers vision, policy, institutional frameworks, regulatory and funding are the underlying drivers for adoption of AI and GIS technologies for rural development.

8. FUTURE DIRECTIONS AND RECOMMENDATIONS

The Scalability of AI and GIS Ecosystems: The Future of Rural Development This would require integration of these technologies across sectors like agriculture, healthcare, education, disaster management, etc. Scaling up GIS-based crop monitoring systems, for example, can lead to better agricultural productivity across the state, while AI-driven predictive analytics can enable optimised resource allocation in sectors such as healthcare and education. Additionally, the development of standardisation and compilation mechanisms, establishing wide initiative and interoperability accomplice along with centralised data welfare can complement and enable smooth data interface (Raheem & Hasan, 2023). The successful adoption of AI and GIS in rural development calls for the collaboration among governments, private enterprises, NGOs and academic institutions. Such partnerships not only promote innovation, but also make certain that technological solutions are affordable for the average decision-maker. It could help us to ensure that solutions implementation is matching local contexts by bridging the gap between technology providers and end-users by NGOs and community-based organizations. Academic institutes can play a role through research in AIGIS applications and training the right talent pool. Collaborative Ecosystem–stakeholders collectively striving towards development objectives, pooling their capabilities.

This highlights some of the key ethical and legal issues that need to be addressed to ensure the responsible use of AI and GIS in rural development. Major worries are data privacy, algorithmic bias and possible displacement of traditional livelihoods. To address these concerns, lawmakers need to enact clear regulations governing data collection, storage, and usage, so individuals' rights are protected; there must be ethical frameworks on AI algorithms, and emphasize transparency, fairness and ensure accountability. Also, there is a need for a balance between the fast-paced growth of technology, and the preservation of traditional knowledge systems and cultural heritage. Addressing these challenges can help stakeholders earn the trust of rural communities and guarantee equitable access to benefits (Pandey & Karnatrak, 2024).

9. CONCLUSION

This infusion of AI into GIS technology will witness a paradigm shift in rural development bringing solutions to age-old challenges faced by rural areas in places like Chhattisgarh. Whether it be enabling communities, creating jobs or enhancing the quality of live of populace, such technologies can be the engine of inclusive and sustainable growth. But realizing this potential must come without robust policy framework, cooperation from stakeholders, and a commitment to address ethical and legal considerations. Thus, by enhancing the scale and synergy of the AI-GIS ecosystem and promoting collaborative partnerships, Chhattisgarh can pave the way for a new chapter in rural development, where no community is left behind in the journey toward progress.

REFERENCES

Arin, I. A., Dhamacora, R., & Aditya, B. (2024, September). The Adoption of Digital Twin Technology to Facilitate Resource and Asset Tracking in Field Service Management. In *2024 International Conference on ICT for Smart Society (ICISS)* (pp. 1-6). IEEE. DOI: 10.1109/ICISS62896.2024.10751066

Aydinoglu, A. C., Bovkir, R., Colkesen, İ., Seyrek, E. C., & Bilgin, M. S. (2024). Developing a mobile application with geo-analytic tools to perform real estate valuation within smart cities. *Survey Review*, 1–19.

Force, I. A. I. T. (2022). A review of the methods, applications, and challenges of adopting artifical intelligence in the property assessment office. *Journal of Property Tax Assessment & Administration*, 19(1), 2.

Griffin, A. L., Reichenbacher, T., Liao, H., Wang, W., & Cao, Y. (2024). Cognitive issues of mobile map design and use. *Journal of Location Based Services*, 18(4), 1–31. DOI: 10.1080/17489725.2024.2371288

Kaldeli, E., Menis-Mastromichalakis, O., Bekiaris, S., Ralli, M., Tzouvaras, V., & Stamou, G. (2021). CrowdHeritage: Crowdsourcing for improving the quality of cultural heritage metadata. *Information (Basel)*, 12(02), 64. DOI: 10.3390/info12020064

Li, J., Xia, H., Qin, Y., Fu, P., Guo, X., Li, R., & Zhao, X. (2022). Web GIS for sustainable education: Towards natural disaster education for high school students. *Sustainability (Basel)*, 14(5), 2694. DOI: 10.3390/su14052694

Li, Y., Zhao, L., Chen, Y., Zhang, N., Fan, H., & Zhang, Z. (2023). 3D LiDAR and multi-technology collaboration for preservation of built heritage in China: A review. *International Journal of Applied Earth Observation and Geoinformation*, 116, 103156. DOI: 10.1016/j.jag.2022.103156

Pandey, K., & Karnatak, H. (2024). Geographic Information. *Advances in Geospatial Technologies for Natural Resource Management*, 41.

Radosevic, N., Duckham, M., Saiedur Rahaman, M., Ho, S., Williams, K., Hashem, T., & Tao, Y. (2023). Spatial data trusts: An emerging governance framework for sharing spatial data. *International Journal of Digital Earth*, 16(1), 1607–1639. DOI: 10.1080/17538947.2023.2200042

Raheem, F. S., & Hasan, A. F. (2023). Enhancing artificial intelligence in advanced database systems for Baghdad's urban transportation management. *Periodicals of Engineering and Natural Sciences*, 11(3), 217–228. DOI: 10.21533/pen.v11i3.3556

Rahimi, M., Malek, M. R., Claramunt, C., & Le Pors, T. (2021). A topology-based graph data model for indoor spatial-social networking. *International Journal of Geographical Information Science*, 35(12), 2517–2539. DOI: 10.1080/13658816.2021.1912349

Sailer, C. (2020). *Enhancing Knowledge, Skills, and Spatial Reasoning through Location-based Mobile Learning* (Doctoral dissertation, ETH Zurich).

Shakhislam, L., Yerlan, I., Duman, A., Ussenov, N., Zhu, K., & Dávid, L. D. (2024). Opportunities and challenges of using geospatial technologies in teaching school geography in Kazakhstan. *International Research in Geographical and Environmental Education*, •••, 1–19. DOI: 10.1080/10382046.2024.2380173

Xia, L. (2022, September). Basic Data Analysis of Smart City Traffic Planning Based on ArcGIS. In *International Conference on Cognitive based Information Processing and Applications* (pp. 289-296). Singapore: Springer Nature Singapore.

Xia, L. (2023, April). Basic Data Analysis of Smart City Traffic Planning Based on ArcGIS Check for updates. In *Proceedings of the 2nd International Conference on Cognitive Based Information Processing and Applications (CIPA 2022):* Volume 1 (Vol. 155, p. 289). Springer Nature. DOI: 10.1007/978-981-19-9373-2_31

Xie, Y., Zhan, N., Zhu, Q., Zhan, J., Guo, Z., Qiao, C., Zhu, J., & Xu, B. (2024). Multimodal data visualization method for digital twin campus construction. *International Journal of Digital Earth*, 17(1), 2431624. DOI: 10.1080/17538947.2024.2431624

Xin, Y., & MacEachren, A. M. (2020). Characterizing traveling fans: A workflow for event-oriented travel pattern analysis using Twitter data. *International Journal of Geographical Information Science*, 34(12), 2497–2516. DOI: 10.1080/13658816.2020.1770259

. Yuan, M. (2024). 15 Frontiers of GIScience. *Remote Sensing Handbook, Volume II: Image Processing, Change Detection, GIS, and Spatial Data Analysis*, 368.

Zhao, Z. (2022, September). Research on Key Technologies of Dynamic Maintenance of Urban and Rural Planning Data Based on Mobile GIS. In *International Conference on Cognitive based Information Processing and Applications* (pp. 721-725). Singapore: Springer Nature Singapore.

Chapter 8
Safety and Security in Autonomous Traffic Systems Management With AI:
Economic Implications of Traffic Optimization in Smart Cities

Anuradha Jain
https://orcid.org/0000-0001-6996-8309
VIPS, India

ABSTRACT

Safety and security are critical pillars in managing autonomous traffic systems with AI, particularly in the context of smart cities. These systems integrate advanced technologies such as IoT, V2X communication, and machine learning to enhance traffic flow, reduce accidents, and optimize urban mobility. However, ensuring the reliability of these systems demands robust safety protocols to address challenges like mixed-traffic scenarios and unexpected system failures. Concurrently, cybersecurity measures are vital to counter threats such as hacking and data breaches that could compromise vehicle and infrastructure integrity. Economically the AI-driven traffic optimization reduces congestion, cuts fuel consumption, and improves productivity by minimizing travel delays. The shift to autonomous traffic systems fosters urban sustainability by decreasing emissions and promoting energy efficiency. While challenges remain in terms of public trust and regulatory frameworks, the economic and environmental benefits position these systems as transformative solutions for the future of urban mobility.

DOI: 10.4018/979-8-3693-9410-6.ch008

1. INTRODUCTION

These advances are leading to systems that can effectively combat contemporary transportation issues such as traffic congestion, road safety, and environmental sustainability. Safety and security provide the basis of managing autonomous traffic systems, vehicle and infrastructure operation and communication network. The efficiency with which these machines operate will minimize the burden on our cities, while at the same time the financial gains made from their operation will also have an immediate impact on our quality of life thanks to a better traffic flow and optimal resource usage (Ji, 2021).

The first aspect of safety in autonomous traffic systems is to design so that accidents cannot happen, and that risks can be managed. The technology behind autonomous vehicles (AVs) is complex, relying on advanced sensors, machine learning algorithms, and real-time data processing to make split-second decisions (Huang et al., 2021). However, ensuring safety in these systems presents problems that must be overcome, including irregularities in road conditions, human errors in mixed-traffic environments (where both human-driven and autonomous vehicles operate), and the malfunction of such systems. AI is equipped with safety measures like collision, risk, trajectory, and environmental perception to help tackle these issues. Predictive systems process vast amounts of data and use predictive modeling to forecast potential threats and corrective actions, allowing for a highly reliable means to power autonomous operations (Tang et al., 2021).

1.1 Evolution of Autonomous Traffic Systems

Along with operational security, this also applies to the security of autonomous traffic systems. These systems rely on inter-linked devices and communication networks, including Vehicle-to-Everything (V2X) communication that enables data transfer between vehicles, infrastructure, and traffic management centers. This connectivity can increase efficiency, but can also open the door for cybersecurity weaknesses. Others pose cybersecurity threats that can intercept data between the vehicle communication systems and the wider cloud network, or hacking into vehicle control systems or tampering with traffic signals, both of which can disrupt operations and create significant public safety emergencies. To secure such systems, we would need to use strong encryption algorithms, make use of blockchain tech for secure interchanges of data storage through decentralized ledgers, and AI will help scan and identifies breaches in real time. Cybersecurity protocols must be implemented to protect sensitive data and ensure public confidence in the reliability of autonomous traffic systems.

Traffic optimization, a key focus of smart city development, has also been at the forefront of AI technologies. Traffic management is a process of analyzing the traffic data in real-time, predicting congestion pods, and rerouting the rides to maximum loss reduction. Deep learning models and neural networks are examples of the type of AI algorithms that can analyze vast datasets from IoT sensors, GPS systems, and traffic cameras to produce accurate forecasts of traffic. On the one hand, the data collected from a wide range of vehicles, including cars, buses, and public transportation, allow these systems to analyze and manage heavy human activities and urban mobility more effectively. Not only does this increased efficiency help daily commuters, it reduces fuel consumption, and lowers greenhouse gas emissions, making cities as a result more environmentally friendly.

The economics of AI-enhanced traffic management are staggering. Through minimized travel distance and enhanced transportation efficiency, self-driving traffic systems help maximize workforce productivity. Commuters lose less time stuck in traffic, thus allowing them took place additional time to skilled and personal activities (Dai et al., 2018). Optimized traffic systems reduce fuel consumption and vehicle maintenance costs, thus providing considerable savings for individuals and companies. By enabling smooth movement of goods and services from one place to another, improved traffic management will attract investments in smart city infrastructure and boost the local economy.

1.2 Role of AI in Transportation: Importance of Safety and Security in Autonomous Systems

Using AI in autonomous traffic systems would also promote sustainable urban areas. As the world urbanizes, with more people living in mega-cities, existing traffic systems become unable to satisfy demand. Autonomous systems can adjust traffic flows to reduce congestion, which would allow urban areas to be more mission-based and preserve the quality of transportation systems as cities grow. These systems contribute to global sustainability objectives and the establishment of green urban environments by lowering emissions and encouraging cleaner modes of transportation.

Challenges to the widespread adoption of autonomous traffic systems one of the major barriers are public confidence in safety and reliability of autonomous systems. Well-publicized collisions have raised questions about the capacity of autonomous vehicles to handle unpredictable and intricate environments on the road. Responding to this skepticism will entail openness surrounding safety protocols, comprehensive testing across a range of scenarios, and maintaining rigorous adherence to regulatory standards (Luettin et al., 2022). To build trust in these systems, policymakers must also grapple with ethical questions, particularly regarding decision-making in unavoidable accidents.

Challenges arise that require regulations and policy frameworks to ensure sustainable growth and development. Governments and international organizations should create rules and regulations for the creation, implementation, and functioning of these systems. But we can lay a not-quite-smooth path forward through collaboration between policymakers, industry stakeholders, and researchers, all seeking to establish standards for safety, security, and innovation. Furthermore, resolving also legal and liability concerns, like understanding who is liable if accidents happen, is vital for encouraging a positive regulatory framework for autonomous traffic systems.

The transformative power of AI is reflected in case studies in cities that have already deployed AI-driven traffic systems. In order to alleviate traffic, the Singapore has utilized AI solution that has reduced the congestion and enhance the commuters experience (Halilaj et al., 2022). Likewise, Stockholm's integration of smart traffic systems has drastically reduced emissions and improved urban mobility. For other cities seeking to implement similar measures, these success stories showcase the economic and environmental advantages of autonomous traffic systems and offer insights into what works.

As the technology continues to foreshadow autonomous traffic lights, machine learning patterns and 5G networks can further optimize traffic light integration. This technology will allow more rapid data movement, refined decision-making, and increased resilience. By integrating renewable energy solutions into traffic infrastructure, we can contribute to sustainability targets and minimize the transportation systems' carbon footprint. AI-based autonomous traffic systems require stringent safety and security management for a successful incorporation into smart cities. With the right framework in place, however, these systems can provide transformative benefits in economic efficiency, sustainability, and urban development addressing safety, cybersecurity, and public trust challenges along the way. With the ongoing evolution of technology, collaboration among governments, industries, and researchers will be crucial in shaping the future of urban mobility and fully unlocking the potential of autonomous traffic systems in building smarter, safer, and more sustainable cities (Xu et al., 2022).

2. AUTONOMOUS TRAFFIC SYSTEMS AND AI INTEGRATION: COMPONENTS OF AUTONOMOUS TRAFFIC SYSTEMS

Smart traffic systems are AI-powered tech systems that can regulate traffic flow, reduce highway congestion, and enhance road safety in combination with autonomous traffic systems. These systems make use of real-time information from other vehicle and infrastructure systems, dynamically changing traffic signalling, optimizing trip planning and efficiently managing urban mobility. Using historical and real-time traffic data, AI algorithms make predictive adjustments that increase the efficiency of traffic flow across the network (Elahi et al., 2020). This could range from detecting anomalies like accidents or heavy congestion and reacting by redirecting traffic or alerting authorities in real time. Integrating with connected infrastructure, these systems allow vehicles including autonomous ones to flow with maximum safety and efficiency.

2.1 Sensors, IoT, and V2X Communication

The first description includes, but is not limited to: Sensors, Internet of Things (IoT) devices, and Vehicle-to-Everything (V2X) communication technology. Cameras, radar, LiDAR and other sensors collect blueprints about vehicle position, speed and surrounding environments. These sensors and centralized traffic systems vehicle-to-x Communication with other vehicles and IoT devices. Connect to a wider platform using V2X, including Vehicle-to-Vehicle (V2V), Vehicle-to-Infrastructure (V2I), and Vehicle-to-Network (V2N) features to facilitate communication with the vehicle environment (Lecue, 2020). Real-time data on traffic conditions, road hazards, and other critical events that this technology enables, is shared between different vehicles, people, networks, and infrastructure; thanks to this vehicle connectivity, drivers have and better situational awareness, and their decisions can be more accurate, whether they are in an autonomous vehicle or a traditional car. V2X technology is a key enabler for cooperative driving strategies, including platooning and adaptive cruise control, which facilitates even better traffic flow and safety.

2.2 Advantages of AI in Traffic Systems: Enhanced Efficiency

Integration of AI in the traffic system can save a lot of time and resources, thus making it much more efficient than before. Utilising trend analysis, AI can prepare for the impending rush hour traffic and tweak the signal timings to best fit the higher traffic. The adjustments are based on live data, guaranteeing minimal disruption and the flow of steady traffic (Kim et al., 2021). AI-driven traffic control also decreases fuel consumption by reducing times of idling and encouraging smoother driving patterns. This efficiency means economic savings for drivers and operators, as well as enhanced mobility for the city overall.

Decrease in Traffic and Pollution

Air pollution and economic inefficiency from road congestion is a huge problem in urban areas. AI-based systems can reduce congestion through rerouting and dynamically identifying alternate paths, and balancing load across the network. These measures reduce stop-and-go traffic patterns, decreasing greenhouse gas emissions and improving air quality. AI also allows for intelligent parking, directing drivers to free spaces, minimizing unnecessary driving. This helps dramatically reduce the environmental characteristics of urban transport systems.

Demand Forecasting Using Machine Learning Models

The field of traffic systems relies heavily on machine learning (ML) to forecast demand. Understanding demand patterns: Supervised learning models can be trained on historical traffic data to predict demand patterns based on the time of day, weather conditions, and local events. Such forecasts will enable city planners and traffic managers to proactively respond to possible bottlenecks by allocating resources or modifying traffic signals as appropriate. In contrast, unsupervised learning models assist in uncovering latent patterns and trends in traffic data leading to more sophisticated and effective management strategies (Guo et al., 2021). Another example is that clustering algorithms can be employed to cluster regions or areas with similar traffic behaviours and to develop enhanced interventions at specific locations.

Traffic Management in Real Time Using Deep Learning

CNNs and RNNs, are the type of deep learning models that ensure the consumption of vast amounts of real-time data in order to deliver the insights. Convolutional Neural Networks (CNNs) excel at processing visual input from traffic cameras to detect accidents and assess traffic density. The sequential nature of traffic data makes RNNs a lucrative and powerful way to predict traffic flow and changes in it. Reinforcement learning (RL) is another strong deep learning method utilized in independent traffic models. Reinforcement Learning (RL) agents learn optimal control policies for traffic signals from their interactions with the environment and based on feedback received after taking actions (Nickel et al., 2015). This allows more adaptive and robust management of complex traffic scenarios, even in unpredictable situations. By going these sophisticated AI models with connected infrastructure and

V2X technology, a fully connected ecosystem for intelligent traffic management, more sustainable and efficient urban mobility becomes possible.

3. SAFETY IN AUTONOMOUS TRAFFIC SYSTEMS

The complexity of the technology and the reality of the environments where the traffic systems will be employed make broad safety standards essential to ensure the safety of autonomous traffic systems. International safety regulatory frameworks and protocols form the bedrock for assessing and certifying autonomous vehicles (AVs). This encompasses system reliability, fault tolerance, and the capacity to function safely in a variety of scenarios. These standards delineate essential safety criteria, including functional safety, cybersecurity, and environmental factors, ensuring that AVs adhere to global best practices.

Beyond the global scope of international regulations, regional frameworks are imperative to the successful harmonization of safety benchmarks. In the U.S. the National Highway Traffic Safety Administration (NHTSA), offers advice to AV makers in its Automated Driving Systems, or ADS, 2.0 and later instruments for a voluntary yet thorough assessment of safety. In Europe, AVs are also required to have intelligent speed assistance and event data recorders under new EU laws, known as the General Safety Regulation, in order to increase accountability and transparency. Autonomous vehicles induce industry benchmarks in addition to regulatory frameworks, facilitating innovation along with safety. Industry-specific standards can come from organizations such as the Institute of Electrical and Electronics Engineers (IEEE) and the International Organization for Standardization (ISO) (Yu et al., 2019). For instance, ISO 26262 defines functional safety for electrical and electronic systems in road vehicles with a focus on risk assessment and mitigation strategies. The significance of these standards lies in the fact that they can help to ensure the AV industry is unified in its approach to ensuring AV safety including stakeholders across the industry.

Even with rigorous safety protocols in place, coordinating self-led traffic systems can also lead to significative risks and malfunctions. Accidents in autonomous traffic present challenges that are specific to the interaction between AVs and their operational environments. However, things like sensor malfunctions, software bugs, and unpredictable outside forces including weather can hinder the effectiveness of AVs (Liu et al., 2023). The reliability and safety of such systems require constant improvement, as highlighted by software bugs in high-impact incidents like Tesla cars using Autopilot; Tesla said it was not using the full automation of driving feature during the crash.

Mixed-traffic environments where autonomous vehicles share the same roads with non-autonomous vehicles create additional complexity. Human drivers can have erratic habits, making unexpected lane changes, ignoring road signs, or providing inconsistent signals that can confuse how AVs respond. AVs encounter these types of dynamic situations on a daily basis, which challenge them to anticipate and react to these situations in real-time, such as on-ramps, roundabouts or pedestrians that didn't get the memo about the crosswalk. In scenarios where the margins for error are razor-thin, a combination of the human decision-making process and the machine logic, coupled with advanced predictive models, is required to curb potential risks.

AI is an essential component of most approaches to autonomous traffic because of the need to be collision-free and risk-averse. Central to these efforts are path-planning algorithms, which allow AMUs to compute an ideal route while bypassing hazards and keeping safe distances from other road occupants. To produce a detailed map of the vehicle's environment, these algorithms use data from a range of sensors, including LiDAR, radar, and cameras. By persistently monitoring this data, AVs can anticipate the paths of nearby entities and take preemptive adjustment measures to avoid possible collisions. Edge cases, rare and borderline scenarios that fall beyond the AV systems training datasets, comprise another layer of risk mitigation complexity. These range from meeting animal crossings on the road, to going through construction sites with too ambiguous signage (Plonnings et al., 2022). Tackling those risks means deploying strict testing regimes, such as simulation-based assessments and on-road experience, which can put AVs through a multitude of potential scenarios.

Real-time hazard detection systems improve upon path planning by allowing for real-time feedback regarding dynamic changes in the environment. AI use cases for public safety Vision systems: AI-based vision systems utilize camera inputs to detect and classify objects like humans, bicycles, and vehicles. These systems leverage deep learning algorithms to analyze visual and spatial data, allowing AV manufacturers to account for unexpected objects or threats in a sub-second timeframe (Kraft et al., 2023). With edge computing, these systems become even more responsive, as they can dramatically reduce the latency associated with processing data and making decisions. Autonomous Traffic Risk mitigation also incorporates V2X (Vehicle-to-everything) communication technologies. Through this mechanism, AVs can communicate with other vehicles, infrastructures, and even pedestrians through the concept of V2X (vehicle to everything) and create a collective environment where situational awareness leads in order to enhance the safety and efficiency of a journey. To illustrate, a V2X enabled vehicle can receive advance notice of a traffic jam or road blockage along its travel route in real-time and is able to alter its path. In the same vein, V2X can enable cooperative maneuvering among AVs, including coordinated lane changes or platooning, to mitigate congestion and maximize traffic throughput (Li et al., 2023).

Beyond solutions of a technical nature, AI-driven safety features are increasingly integrating ethical decision-making frameworks. The frameworks treat scenarios in which AVs have to prioritize outcomes (e.g., whether to minimize harm to passengers or pedestrians in a crash that is unavoidable). But the development of transparent and explainable AI (XAI) systems ensures that these decisions are understandable and accountable, thereby promoting public trust in the autonomous traffic systems.

4. SECURITY IN AUTONOMOUS TRAFFIC SYSTEMS

Security is among the most crucial pillars for the development and deployment of autonomous traffic since vulnerabilities in these systems can cause catastrophic events. Due to the interconnected nature of vehicles, infrastructure, and the broader traffic ecosystem, there are several cybersecurity challenges in autonomous systems (Schier et al., 2023). AVs are dependent on complex hardware and software integrations (sensors, communication modules, cloud-based services) that expose the vehicle to attack vectors. Cyberattacks on autonomous systems and traffic infrastructure range broadly from hacking into vehicle systems to the large-scale interruption of the traffic management networks. Breach could enable hackers to take control of an autonomous vehicle, potentially causing crashes or disrupting traffic flow. A notable example is the widely publicised successful demonstration of vehicle hacking in which researchers were able to remotely manipulate a connected car's braking and steering systems, providing concrete evidence that these IoT vulnerabilities can be exploited in the real world. Cyberattacks on smart traffic infrastructure traffic lights, surveillance cameras and road sensors could stall urban mobility and compromise public safety (Mahon et al., 2020).

Such systems produce and handle massive volumes of sensitive data, ranging from real-time positions of the vehicles to identification data of the passengers and diagnostics of the (software) systems inside the vehicles. Any breach of this data jeopardizes privacy and opens the door for criminal usage, like identity theft and/or surveillance. Moreover, data tampering can result in AVs making incorrect decisions, such as misreading road conditions or traffic signals, with potentially fatal consequences. Mitigating these risks necessitates strong cybersecurity protocols to safeguard data integrity, confidentiality and availability throughout the autonomous landscape.

4.1 Usage: Blockchain and Encryption

Advanced technologies such as blockchain and encryption are being used in autonomous traffic systems to counter cybersecurity threats. The decentralized and immutable ledger offered by BlockChain presents an effective way to provide security of vehicle to infrastructure communication. Blockchain keeps all transactions and messages secure and immutable by providing a secure channel for data exchange due to its trustless system. Blockchain technology has the potential to enhance V2V and V2I communications by providing a verifiable framework for data sharing, ensuring that critical information is accurate and trustworthy, helping to protect against false information spread.

Blockchain technology enables secure data sharing between different stakeholders in the autonomous ecosystem as well. This is especially true for nervous situations where data is being aggregated from multiple sources as car manufacturers, traffic authorities, insurance companies, etc. These entities benefit from using a blockchain-based system to share data in a secure manner to ensure transparency and accountability (Xiao et al., 2023). In cases in which an accident transpires, for instance, blockchain can offer an indisputable chain of evidence, which can serve to determine liability (and ensure evidence is not altered). Autonomous traffic systems involve various mobile devices, and encryption is a key strategy to secure the communication channels of these mobile devices. End-to-end encryption enables data that is exchanged between vehicles, infrastructure, and cloud servers to be visible only to authorized entities. Sensitive data including navigation routes, vehicle diagnostics, and passenger details is encrypted to prevent interception and unauthorized access from attackers using advanced encryption algorithms like RSA and AES.

Effective encryption implementation in autonomous systems also requires secure key management. Explorations into cutting-edge methods like quantum key distribution (QKD) allow exchanges to happen with an impenetrable layer of protection, able to sense if someone is listening in by utilizing quantum mechanics principles. With the use of blockchain and encryption, autonomous traffic systems can secure their level of security and ensure their operations are resilient and trustworthy.

4.2 AI-Driven Threat Detection

Artificial Intelligence (AI) is at the heart of detecting and counteracting cyberattacks on autonomous traffic systems in real time. Moreover, traditional security methods like firewalls and intrusion detection systems struggle to cope with the dynamic and complex nature of cyber threats to AVs and their ecosystems. In contrast, AI-based threat detection systems can process massive datasets, recognize patterns indicative of malicious activity, and react to threats with a degree of speed and precision never before seen (Liu et al., 2020).

Anomaly detection is one of the main uses of AI in cybersecurity. Machine learning algorithms enable AI systems to develop baseline patterns of normal behavior for autonomous vehicles and infrastructure. If any activity diverges from the usual standard like a sudden spike in traffic on the network or access attempts from an unexpected geography alerts are triggered and the security team can investigate and respond to potential throat in real time. By training AI models on large datasets of known cyberattacks, it can also use these systems to identify any new and emerging attack vectors, making them crucial in the combating of sophisticated and evolving threat actors.

Real-time monitoring and protection of communication channels is another key application of AI. For instance, this is a crucial aspect when it comes to autonomous traffic systems that lean on V2V and V2I communication for sharing information related to road conditions, traffic signals, and potential hazards. AI-powered systems can analyze these communications for anomalies, and can correct any issues that they find such as messages that have been spoofed, or unauthorized data that has been injected to keep the network secure and functioning properly (Liu et al., 2020). For example, if an AI detects a fake traffic signal broadcast to create confusion between AVs, the AI can detect the source of that signal, and alert AVs to ignore the signal. Autonomous traffic systems security can be further reinforced through AI-powered predictive analytics, which estimate potential vulnerabilities and attacks. AI can study historical data to find patterns, allowing it to predict possible targets and attack methods, which allows enterprises to take action to bolster defenses before attack even happens. For instance, if a specific type of sensor or software module has been flagged as a frequent entry point for hackers, AI can prioritize its monitoring and suggest updates or patches to reduce the risk. AI can also be integrated into cybersecurity frameworks for incident response and recovery. AI technologies are capable of automating critical components of incident response; for example, in actions like isolating compromised systems, restoring backups, and applying patches significantly decreasing downtime and mitigating the broader impact of cyberattacks. AI can also offer insights into the underlying causes of incidents, allowing organizations to re-evaluate and strengthen their security strategies to avoid similar issues in the future.

5. ECONOMIC IMPLICATIONS OF TRAFFIC OPTIMIZATION: COST-BENEFIT ANALYSIS OF AUTONOMOUS TRAFFIC SYSTEMS

Autonomous traffic systems can impact the economic dimensions of urban mobility significantly, with not only the aspects of safety in mind but also with a view on how, in the future, urban mobility has the potential to change economy itself. Autonomous technologies responsible for traffic optimization can yield significant economic benefits, but they also come with costs that need to be thoroughly assessed (Bonatti et al., 2019). Cost-benefit Analysis of All These SystemsPresenting potential costs, revenues, investments, savings and long-term effects that could disrupt the transportation economic future. Avoid Traffic congestion with Autonomous system is one of the reasons of this economy. Traffic jams already cost global economies billions of dollars of lost productivity, fuel wasted and delays every year. Leveraging advanced sensors and communication modules, as well as real-time data analytics, autonomous systems can manage vehicle speeds, coordinate intersections, and dynamically reroute traffic to optimize flow and avoid congestion-related issues. As a result, these measures considerably minimize unproductive and fuel use save, which is translating into reduced operational expenses for private and commercial cars. For example, research shows that the extensive use of AVs could lead to fuel consumption declines of up to 25% because of smoother acceleration and deceleration patterns (Zheng et al., 2022). This cut helps individual drivers and has a cumulative impact on industries that rely on transportation, like logistics and e-commerce, which often operate on skinny profit margins. An efficient traffic system also extends the lifespan of infrastructure, as less congestion in the area leads to less wear and tear on points of entry, and less time spent doing maintenance, so that public money can be reinvested into other avenues.

Much infrastructure and technology must be constructed and purchased to facilitate autonomous traffic systems. From smart traffic lights and advanced sensors, to connected vehicle networks and centralized traffic management hubs, many components facilitate an optimized system. In order to pursue these goals, governments and private stakeholders have to invest significant resources in the research, development, deployment, and maintenance of these technologies. Although these expenses seem high at first, they really are investments that will pay economic dividends in the long term. Indeed, smart intersections could potentially reduce accident costs by avoiding accidents through more accurate traffic timing. Likewise, the implementation of vehicle-to-everything (V2X) communication systems allows for seamless data transfer between vehicles to the infrastructure, streamlining travel more effectively and safely. Governments may also benefit from increased revenue streams through new business models, such as mobility-as-a-service (MaaS), where conventional car ownership would be replaced by shared autonomous fleets (Werner et al., 2021).

In other words, traffic optimization leads to improved efficiency, which in turn has very real economic productivity effects. In crowded cities, commutes can mean lost hours of work, at a great cost to both employers and employees. Autonomous traffic systems by decreasing travel time and increasing predictability allow workers to be on the road less and engaged in productive activities more. For businesses that depend on just-in-time logistics, the accurate and timely delivery of goods through AI can be particularly significant, as delays in supply chains can lead to broken operations and lost income (Buchgeher et al., 2021). Self-driving traffic systems open exciting possibilities for restructuring urban design and land usage (Raghav et al., 2024). Less congestion and parking demand may give cities the opportunity to repurpose spaces for commercial / residential / recreational, stimulating the economy as well as enhancing the quality of life (Kaushik et al., 2024). The overall economic value of such urban transformations will likely far exceed the initial implementation costs as smart cities continue to flourish in the future (Singh et al., 2024).

6. CONCLUDING REMARKS AND FUTURE ASPECTS

AI is a veritable game-changer, leading to a safer and more robust transportation system, as well as a cornerstone for effective and resilient smart cities. These systems greatly improve public safety and economic development by minimizing accidents, improving live threat detection, and optimizing traffic flow (Sun et al., 2023). Cutting-edge technologies such as blockchain, encryption, and AI-supported risk mitigation protect critical infrastructure while facilitating secure communication between vehicles and traffic management centers. The resulting boost in transportation efficiency decreases costs related to congestion and fuel consumption, while also releasing productivity gains that can spur economic recovery in cities. The potential for AI in autonomous traffic management is enormous moving forward. Looking ahead, we may also see the incorporation of quantum computing to facilitate high-speed, accurate decision-making, the advancement of V2X communication for a universal communication network, as well as the use of predictive analytics to combat risks to safety and security before they occur. Moreover, autonomous traffic systems can also support sustainable urban initiatives by enabling the increased use of shared and electric vehicles, thereby decreasing carbon emissions and encouraging greener modes of transport

It is necessary to tackle challenges such as regulation, public trust, and equitable access to technology: challenges that must be addressed on the long road to reach these advances. Spending on research and development, infrastructure, and workforce impacted by reskilling to be able to overcome these challenges. Evolution of Smart Cities: Building Safe Resilient Society through Autonomous Traffic Systems. This will enable them to realise the complete potential of AI-enabled autonomous traffic and create better, safer & secure economically rich ambiance.

REFERENCES

Bonatti, P. A., Decker, S., Polleres, A., & Presutti, V. (2019). AIs: New directions for knowledge representation on the semantic web (dagstuhl seminar 18371). *Dagstuhl Reports*, 8(9).

Buchgeher, G., Gabauer, D., Martinez-Gil, J., & Ehrlinger, L. (2021). AIs in manufacturing and production: A systematic literature review. *IEEE Access : Practical Innovations, Open Solutions*, 9, 55537–55554. DOI: 10.1109/ACCESS.2021.3070395

Büchner, M., Zürn, J., Todoran, I. G., Valada, A., & Burgard, W. (2023). Learning and aggregating lane MLs for urban automated driving. In *Proceedings of the IEEE/CVF Conference on Computer Vision and Pattern Recognition* (pp. 13415-13424).

Cai, P., Wang, H., Sun, Y., & Liu, M. (2022). DQ-GAT: Towards safe and efficient autonomous driving with deep Q-learning and ML attention networks. *IEEE Transactions on Intelligent Transportation Systems*, 23(11), 21102–21112. DOI: 10.1109/TITS.2022.3184990

Dai, S., Liang, Y., Liu, S., Wang, Y., Shao, W., Lin, X., & Feng, X. (2018, March). Learning entity and relation embeddings with entity description for AI completion. In *2018 2nd International Conference on Artificial Intelligence: Technologies and Applications (ICAITA 2018)* (pp. 194-197). Atlantis Press.

Dong, X. L., He, X., Kan, A., Li, X., Liang, Y., Ma, J., & Han, J. (2020, August). Autoknow: Self-driving knowledge collection for products of thousands of types. In *Proceedings of the 26th ACM SIGKDD International Conference on Knowledge Discovery & Data Mining* (pp. 2724-2734). DOI: 10.1145/3394486.3403323

Elahi, M. F., Luo, X., & Tian, R. (2020, July). A framework for modeling AIs via processing natural descriptions of vehicle-pedestrian interactions. In *International Conference on Human-Computer Interaction* (pp. 40-50). Cham: Springer International Publishing.

Guo, R., Hong, Z., & Xue, X. (2021). *Reward Function Design via Human AI and Inverse Reinforcement Learning for Intelligent Driving* (No. 2021-01-0180). SAE Technical Paper.

Halilaj, L., Dindorkar, I., Lüttin, J., & Rothermel, S. (2021). A AI-based approach for situation comprehension in driving scenarios. *The Semantic Web: 18th International Conference, ESWC 2021, Virtual Event, June 6–10, 2021Proceedings*, 18, 699–716.

Halilaj, L., Luettin, J., Henson, C., & Monka, S. (2022, September). AIs for automated driving. In *2022 IEEE Fifth International Conference on Artificial Intelligence and Knowledge Engineering (AIKE)* (pp. 98-105). IEEE.

Halilaj, L., Luettin, J., Monka, S., Henson, C., & Schmid, S. (2023). AI-Based Integration of Autonomous Driving Datasets. *International Journal of Semantic Computing*, 17(02), 249–271. DOI: 10.1142/S1793351X23600048

Holzinger, A., Saranti, A., Hauschild, A. C., Beinecke, J., Heider, D., Roettger, R., . . . Pfeifer, B. (2023, August). Human-in-the-loop integration with domain-AIs for explainable federated deep learning. In *International Cross-Domain Conference for Machine Learning and Knowledge Extraction* (pp. 45-64). Cham: Springer Nature Switzerland. DOI: 10.1007/978-3-031-40837-3_4

Hu, X., Liu, Y., Tang, B., Yan, J., & Chen, L. (2023). Learning dynamic ML for overtaking strategy in autonomous driving. *IEEE Transactions on Intelligent Transportation Systems*, 24(11), 11921–11933. DOI: 10.1109/TITS.2023.3287223

Huang, H., Sun, L., Du, B., Liu, C., Lv, W., & Xiong, H. (2021, August). Representation learning on AIs for node importance estimation. In *Proceedings of the 27th ACM SIGKDD Conference on Knowledge Discovery & Data Mining* (pp. 646-655).

Ji, S., Pan, S., Cambria, E., Marttinen, P., & Philip, S. Y. (2021). A survey on AIs: Representation, acquisition, and applications. *IEEE Transactions on Neural Networks and Learning Systems*, 33(2), 494–514. DOI: 10.1109/TNNLS.2021.3070843 PMID: 33900922

Kaushik, T. K., Singh, A., Raghav, A., Singh, B., Arora, M. K., & Lal, S. (2024). Untapping Artificial Intelligence in Environmental and Sustainable Progression: Projecting Legal and Ethical Considerations. In B. Singh, C. Kaunert, K. Vig, & S. Dutta (Eds.), *Maintaining a Sustainable World in the Nexus of Environmental Science and AI* (pp. 109-130). IGI Global. DOI: 10.4018/979-8-3693-6336-2.ch005

Kim, J. E., Henson, C., Huang, K., Tran, T. A., & Lin, W. Y. (2021). Accelerating road sign ground truth construction with AI and machine learning. In *Intelligent Computing:Proceedings of the 2021 Computing Conference,* Volume 2 (pp. 325-340). Springer International Publishing.

Kraft, M., Bai, J., Mosbach, S., Taylor, C., Karan, D., Lee, K. F., ... Lapkin, A. (2023). From Platform to AI: Distributed Self-Driving Laboratories.

Lecue, F. (2020). On the role of AIs in explainable AI. *Semantic Web*, 11(1), 41–51. DOI: 10.3233/SW-190374

Li, X., Liu, J., Li, J., Yu, W., Cao, Z., Qiu, S., . . . Jiao, X. (2023, August). ML Structure-Based Implicit Risk Reasoning for Long-Tail Scenarios of Automated Driving. In *2023 4th International Conference on Big Data, Artificial Intelligence and Internet of Things Engineering (ICBAIE)* (pp. 415-420). IEEE.

Liu, H., Li, Y., Hong, R., Li, Z., Li, M., Pan, W., Glowacz, A., & He, H. (2020). AI analysis and visualization of research trends on driver behavior. *Journal of Intelligent & Fuzzy Systems*, 38(1), 495–511. DOI: 10.3233/JIFS-179424

Liu, J., Zhang, X., Li, Y., Wang, J., & Kim, H. J. (2019). Deep learning-based reasoning with multi-ontology for IoT applications. *IEEE Access : Practical Innovations, Open Solutions*, 7, 124688–124701. DOI: 10.1109/ACCESS.2019.2937353

Liu, Q., Li, X., Tang, Y., Gao, X., Yang, F., & Li, Z. (2023). ML Reinforcement Learning-Based Decision-Making Technology for Connected and Autonomous Vehicles: Framework, Review, and Future Trends. *Sensors (Basel)*, 23(19), 8229. DOI: 10.3390/s23198229 PMID: 37837063

Luettin, J., Monka, S., Henson, C., & Halilaj, L. (2022, November). A survey on AI-based methods for automated driving. In *Iberoamerican AIs and Semantic Web Conference* (pp. 16–31). Springer International Publishing.

Mahon, L., Giunchiglia, E., Li, B., & Lukasiewicz, T. (2020, December). AI extraction from videos. In *2020 19th IEEE International Conference on Machine Learning and Applications (ICMLA)* (pp. 25-32). IEEE.

Nickel, M., Murphy, K., Tresp, V., & Gabrilovich, E. (2015). A review of relational machine learning for AIs. *Proceedings of the IEEE*, 104(1), 11–33. DOI: 10.1109/JPROC.2015.2483592

Ploennigs, J., Semertzidis, K., Lorenzi, F., & Mihindukulasooriya, N. (2022, October). Scaling AIs for automating AI of digital twins. In *International Semantic Web Conference* (pp. 810-826). Cham: Springer International Publishing.

Raghav, A., Singh, B., Jermsittiparsert, K., Raghav, R., & Yadav, U. (2024). Artificial Intelligence in Environmental and Climate Changes: A Sustainable Future. In Singh, B., Kaunert, C., Vig, K., & Dutta, S. (Eds.), *Maintaining a Sustainable World in the Nexus of Environmental Science and AI* (pp. 485–506). IGI Global. DOI: 10.4018/979-8-3693-6336-2.ch019

Schier, M., Reinders, C., & Rosenhahn, B. (2023). Deep reinforcement learning for autonomous driving using high-level heterogeneous ML representations. In *International Conference on Robotics and Automation (ICRA)*.

Sharma, A., & Singh, B. (2022). Measuring Impact of E-commerce on Small Scale Business: A Systematic Review. *Journal of Corporate Governance and International Business Law*, 5(1).

Singh, B. (2022). COVID-19 Pandemic and Public Healthcare: Endless Downward Spiral or Solution via Rapid Legal and Health Services Implementation with Patient Monitoring Program. *Justice and Law Bulletin*, 1(1), 1–7.

Singh, B., Kaunert, C., Malviya, R., Lal, S., & Arora, M. K. (2024). Scrutinizing Consumer Sentiment on Social Media and Data-Driven Decisions for Business Insights: Fusion of Artificial Intelligence (AI) and Business Intelligence (BI) Foster Sustainable Growth. In Natarajan, A., Galety, M., Iwendi, C., Das, D., & Shankar, A. (Eds.), *Intersection of AI and Business Intelligence in Data-Driven Decision-Making* (pp. 183–210). IGI Global. DOI: 10.4018/979-8-3693-5288-5.ch007

Sun, R., Lingrand, D., & Precioso, F. (2023). Exploring the Road ML in Trajectory Forecasting for Autonomous Driving. In *Proceedings of the IEEE/CVF International Conference on Computer Vision* (pp. 71-80).

Tang, X., Sun, T., Zhu, R., & Wang, S. (2021, January). CKG: dynamic representation based on context and AI. In *2020 25th International Conference on Pattern Recognition (ICPR)* (pp. 2889-2895). IEEE.

Werner, S., Rettinger, A., Halilaj, L., & Lüttin, J. (2021). RETRA: Recurrent transformers for learning temporally contextualized AI embeddings. *The Semantic Web: 18th* Springer International Publishing.]. *Proceedings*, 18, 425–440.

Xiao, D., Dianati, M., Geiger, W. G., & Woodman, R. (2023). Review of ML-based hazardous event detection methods for autonomous driving systems. *IEEE Transactions on Intelligent Transportation Systems*, 24(5), 4697–4715. DOI: 10.1109/TITS.2023.3240104

Xu, W., Feng, L., & Ma, J. (2022). Understanding the domain of driving distraction with AIs. *PLoS One*, 17(12), e0278822. DOI: 10.1371/journal.pone.0278822 PMID: 36490240

Yu, C., Wang, X., Xu, X., Zhang, M., Ge, H., Ren, J., Sun, L., Chen, B., & Tan, G. (2019). Distributed multiagent coordinated learning for autonomous driving in highways based on dynamic coordination MLs. *IEEE Transactions on Intelligent Transportation Systems*, 21(2), 735–748. DOI: 10.1109/TITS.2019.2893683

Zheng, Z., Zhou, B., Zhou, D., Soylu, A., & Kharlamov, E. (2022, October). Executable AI for transparent machine learning in welding monitoring at bosch. In *Proceedings of the 31st ACM International Conference on Information & Knowledge Management* (pp. 5102-5103).

Chapter 9
Smart Cities Ahead:
Policy and Regulatory Strategies for Machine Learning and Robotics in Urban Development

Sahil Lal
https://orcid.org/0000-0001-9827-3717
Sharda University, India

Kittisak Wongmahesak
https://orcid.org/0009-0000-2129-4691
North Bangkok University, Thailand

Manmeet Kaur Arora
https://orcid.org/0009-0002-5071-117X
Sharda University, India

Christian Kaunert
https://orcid.org/0000-0002-4493-2235
Dublin City University, Ireland

Bhupinder Singh
https://orcid.org/0009-0006-4779-2553
Sharda University, India

ABSTRACT

This chapter examines how machine learning (ML) algorithms and robotics could help redesign cities for a future through policies and regulatory approaches. With rapid urbanization, environmental sustainability, and resource constraints emerging as pressing challenges for cities, the adoption of technology-enabled solutions can significantly improve the management of urbanization, resulting in an improved

DOI: 10.4018/979-8-3693-9410-6.ch009

quality of life in cities. The chapter surveys existing policies on smart city initiatives, analyses regulatory challenges around data privacy and algorithmic bias, and compares approaches to smart city governance across countries. If properly implemented, ML and robotics can contribute to effective allocation of resources, service delivery and citizen engagement, as illustrated with successful case studies. It also sheds light on the lessons learned from failed initiatives, pointing to the importance of inclusive policymaking and strong data governance frameworks.

1. INTRODUCTION

Global urbanization and rapid urban growth gave rise to the notion of smart cities. A smart city is an urban space that uses the latest technology and data-driven solutions to enhance citizens' quality of life without compromising sustainability. This transformation is premised on the adoption of information and communication technologies (ICT), internet of things (IOT) and distinctive urban design (Allam & Jones, 2020). It conclude with recommendations for policymakers, academia, and industry stakeholders. In conclusion this chapter seeks to provide an important contribution to the reflection in the design of advanced strategies that leverage the potential of ML and robotics in urban transformation, considering its complexity at his core. This study looks at the intersection of these technologies and urban governance and policy, with a particular focus on policy and regulatory dimensions that incentivize their uptake. (Angelidou et al., 2021).

1.1 Background of Smart Cities

Smart city can mean lots of things, but at its most basic, it can be interpreted as urban spaces utilizing digital technology to gather data and be more efficient in services but most of all generate well-managed cities. Smart-city strategies have origins that reach back as far as the 1960s and 1970s when budget-constrained urban planners used computer databases and aerial photography to design city and departments and distribute resources (Anthopoulos & Reddick, 2020). Over the decades, there have been three generations of Smart Cities, from Smart City 1.0 driven by technology providers, to Smart City 2.0 slated by the local government, and finally Smart City 3.0 addressing citizen co-creation and participation in the context of urban settings.

Developing sustainable urban settings has become a global goal in recent years, fueled by rising population densities and environmental concerns. Almost two-thirds of the world's population will live in cities by 2050, according to the United Nations, and innovative solutions are needed to manage urban infrastructure effectively. Smart cities provide solutions to these problems by the implementation of advanced technologies across several sectors including transportation, energy management, waste disposal, and public safety. The ultimate objective is to develop interlinked systems that streamline operations and promote economic development and citizen welfare.

There are significant criticisms of the smart cities concept. Critics say that too many smart city initiatives are too tech-focused at the expense of community needs or result in increased surveillance and inequity among residents. For the data is a one yet successful examples show that as long as effective, smart city strategies can greatly improve urban living conditions through the provision of real-time data insights that help and assist decision-making processes.

1.2 Importance of Machine Learning and Robotics in Urban Development

Contents[show]Machine learning (ML) and robotics are among the highest-level technologies shaping smart cities. They allow municipalities to tap into the tons of data coming from everything from traffic sensors to social media posts to make better policy and operational decisions. Based on past patterns, machine learning algorithms can predict trends for the future, use resources efficiently, and improve service delivery in various areas like transportation management, energy consumption, and public health.

For example, by analyzing real-time traffic data and adjusting the length of signal timings, ML applications enable intelligent traffic management systems to alleviate congestion. Robotics is equally important for automating functions like waste collection and maintenance of public spaces, increasing efficiency and relieving humans from simpler tasks to take care of more challenging duties (Bibri, 2021). Such technologies promote a more responsive city that can respond to changing conditions. Additionally, machine learning specializes in personalizing public services based on user data to ensure that citizens feel more engaged within their respective communities. Using information acquired through analyzing how users interact and feedback, municipalities can adjust services, for example, public transport time tables or emergency notifications, to better serve the user and build satisfaction. Such strategic planning and decision-making help local governments become increasingly data-driven.

ML and robotics are not only focused on making our operations better in a business sense, but they are also the key part of bringing solutions to global issues, for instance global warming (Bolívar, 2021). The fusion of smart technologies within smart city initiatives helps in sustainability as they optimize energy consumption, minimize outputs, and assist in improved air quality monitoring. The use of machine learning and robotics will proactively support the sustainability targets as increasingly dense populations pack into our cities.

1.3 Objectives of the Research

The objective of this research paper is therefore to identify and discuss the policy and regulatory levers necessary to design adequate urban development frameworks that ensure effective integration of ML and R. The objectives are threefold:

I. **Analyse Current Policies:** The first goal is to analyse current policies on smart city developments worldwide. This includes identifying best practices and gaps in current regulations that can stifle innovation or equitable access to smart city benefits.
II. **To Assess Impacts of Technology:** The second one is to evaluate how do existing machine learning and robotics applications in urban planning and management. The emphasis of this assessment will be on the influence they have on service delivery, citizen engagement, environmental sustainability and urban resilience in general.
III. **To Make Recommendations:** In the end, this study will make recommendations for policymaking, drawing from the current policies analysis and the technological impacts analysis. To ensure that the identified areas foster innovation while ensuring inclusivity and equity in smart city technology access, our recommendations will focus on these initiatives as positive and supportive regulations.

To conclude, while the integration of technology in urban environments holds enormous potential for enhancing quality of life, it must be pursued with caution and foresight ensuring that we build cities of the future that truly meet the needs and expectations of their citizens (Cavada et al., 2020). This discussion about the context of smart cities, the importance of new technologies in building better cities, and the proposed research agenda oriented towards informing a policy agenda provides the framing necessary to discuss how smart cities should be governed.

2. LITERATURE REVIEW

In recent years, the idea of smart cities has gained momentum as municipalities find themselves facing the challenges of rapid population growth, environmental sustainability, and a need to improve the quality of life. The review covers the theoretical foundations of smart cities, literature related to the use of machine learning techniques in contemporary urban planning, the contribution of robotics to urban governance, and the role of existing policy frameworks in shaping smart city planning.

2.1 Overview of Smart City Concepts

Smart Cities are Urbanized areas that make use of digital technologies to improve their overall performance and well-being. Info & Communication Technology (ICT) and (IoT) to aid this transformation by analysing the data collected by the devices, thus allowing for better/real time decision-making (Cheng & Jin, 2020). A smart city, as defined by the European Commission, includes efficient volatile urban transportation networks with smart public transportation, smart urban planning and upgraded waste disposal facilities, interactive city administration, and a firm agenda for citizen engagement. Smart cities, in general have evolved from the 1990s and as cities have started using technology to solve the challenges faced by them. These two technologies were initially used more behind the scenes, but as people became more aware of them, the goal became how to start engaging the citizens of the smart city in its ecosystem. A connected environment, data-driven decision-making, sustainable infrastructure and enhanced public services are the hallmarks of smart cities today (Cocchia & Dameri, 2021).

Initiatives like the Smart Cities Mission launched in India in 2015 focuses on the making of inclusive and sustainable cities by improving the infrastructure in Tier 1 and Tier 2 cities. This initiative is reflective of a global movement supporting the retrofitting of existing urban fabric as opposed to new greenfield developments. Smart city infrastructure is offered on the premises of essential aspects of the sheer concept: stable electricity supply, waste disposal systems, affordable housing, and strong ground connection. In this way, with smart cities, the idea is to create independent ecosystems for urban sectors of life that cater to the needs of a wide range of users, contributing to economic development whilst also ensuring overall sustainability. Smart city initiatives must therefore find this balance, ensuring that technology is a tool for social equity and environmental responsibility.

2.2 Current Applications of Machine Learning in Urban Planning

Machine learning (ML) is being applied more frequently to urban planning to interpret large corpuses of data from sources including traffic sensors, social media activity, and environmental observational systems. Such applications empower city planners to make better decisions that improve the quality-of-service delivery and improve resource allocation. Traffic management is one of the significant applications of ML in urban planning. Machine learning algorithms use live traffic data to predict traffic congestion trends and recommend desired routing for public transportation systems. For one, cities like Los Angeles have implemented machine learning traffic signal optimization systems that automatically adjusts signal cycles based on live video data, which drastically reduces travel times (and emissions) in general (Dameri & Rosenthal-Sabroux, 2020).

Moreover, predictive maintenance on urban infrastructure uses machine learning. ML models can be applied on historical data regarding infrastructure performance and can predict when maintenance work is needed for public transport systems, bridges, and roads. Not only does this approach increase the life of the assets, but it also decreases interruptions in day to day life. Machine learning also has a critical impact on environment monitoring. Air Quality Data: ML Algorithms Becomes the Leading Way For Cities To Read Air Quality Data Collected By Sensors In The City. This analysis can help identify pollution hotspots and inform policy decisions that lead to an improvement of air quality standards. Machine learning also helps in optimizing energy consumption in buildings by looking at usage patterns and recommending energy-saving approaches. The use of machine learning applications in modern city or urban planning not only improves the efficiency of operations but also enables enterprises to adopt sustainability practices (Deakin, 2020). These technologies are important for solving complex urban problems as cities increasingly transform into intelligent systems.

2.3 Role of Robotics in Urban Management

Another technology liberating city management practices is robotics. Utilization of robotic systems can lead to improved efficiency in fields like waste management, public safety and infrastructure. So, in the domain of waste management, expectedly, autonomous robotics are used along with sensory intelligence to automate garbage collection processes. IoT sensors of waste bins can detect the fill levels, which makes these processes autonomous and automated as these robots reach the waste bins that need to be cleared by navigating through the streets. This not only helps in optimizing collection routes but also minimizes logics costs in terms of collection methods.

Robots are vitalizing public safety in smart cities as well. That can include monitoring public spaces for potential security threats or emergencies using surveillance drones outfitted with cameras and sensors; some municipalities have deployed such systems. In critical incidents, the use of drones can help law enforcement agencies tap into real-time data for faster response times (Dener & Sutherland, 2021).

Additionally, robotics can aid infrastructure maintenance by inspecting bridges or tunnels with drones or robotic arms fitted with imaging technology. It enhances safety during inspections and minimizes labour costs for manual assessments. Not only does the integration of robotics in urban management improve operational efficiency, enabling service providers to respond promptly to the needs and queries of the residents, but it also retains and increases service quality. Robotics is likely to have a bigger hand in city management as technology develops.

2.4 Policy Frameworks in Existing Smart City Initiatives

Smart city deployments typically prioritize policy frameworks that delineate guidance for the technology deployed as well as governance structures and communication channels that constitute successful projects. The balancing act is regulatory compliance without stifling innovation and incorporating equitable access to smart cities' benefits. On one hand, they should allow innovation to flourish and be responsive to the regulatory climate; on the other hand, governments should ensure equitable access to the benefits provided by smart cities.

Existing policy frameworks to bring together government agencies, the private sector, and community stakeholders (Díaz-Díaz et al., 2020). Many smart city initiatives promote a philosophy of open data and share operational data with citizens to promote transparency and community engagement. Such participatory approach reinforces the trust of citizens with local governments and facilitates engaged citizenship.

In addition to fostering collaboration, effective policy frameworks should tackle privacy and security challenges posed by extensive data collection in smart cities. To this end, we call upon policymakers to implement frameworks that will protect citizens from the storms of the virtual world, while enabling the responsible flow of data across and within stakeholders. It also means having to manage cybersecurity threats and the secure nature of the sensitive data being captured via these cutting-edge technologies from smart cities. Moreover, policies also needs to be inclusive, in the sense that every persons are able to benefit from the smart city services, irrespective of socio-economic status. This may be through targeting capital in under-served communities, or tackling a digital divide access through improved internet availability. As cities worldwide continue to evolve into sophisticated entities capable of addressing contemporary challenges through innovative means, it is imperative that legislators establish parameters for technology deployment that enhance the lives of individuals. (Eremia et al., 2020).

3. POLICY AND REGULATORY FRAMEWORKS

Strong policy and regulatory frameworks are increasingly driving how smart cities are developed. They also are essential for the ethical integration of advanced technologies such as machine learning and robotics in urban planning and governance (Giffinger & Gudrun, 2020).

3.1 Overview of Current Policies Affecting Smart Cities

Due to the varying structures of governance, economic conditions and urban challenges across continents, smart city policies vary widely around the globe. Many countries have initiated smart city programs, an effort led by national governments, harnessing modern technology to improve urban infrastructure, sustainability, and the citizens quality of life.

One interesting national policy initiative in India for transforming cities into life and work-designated hubs of innovation and economic dynamism is the Smart Cities Mission. Launched in 2015, this initiative aims to build 100 cities through a mix of public-private partnerships (PPP) and large-scale government investment. Also The Great Indian Development Story: The Mission Core Infrastructure Core infrastructure development will be a focus of the mission suitable water supply systems and waste management practices and developing sustainable housing solutions (Gil-Garcia et al., 2020).

What is emerging are the kind of federal policies that we need to further enable smart cities to take root, including funding and technical assistance. So far, the U.S. Department of Transportation has run multiple projects to facilitate the adoption of intelligent transportation system (ITS) with features such as analytics and machine learning capabilities to manage the traffic. In addition to CCTV, other measures to protect critical infrastructure from potential risk posed by technology are also raising concern from the feds.

European countries have set the pace in providing the foundation for development of smart cities. The European Commission, similarly, has introduced initiatives like the European Urban Initiative (EUI) to help cities implement good practices in sustainable urban development. In addition to this, the EUI fosters inter-municipal collaboration with private actors and civilians to boost innovation while ensuring compliance with environmental legislation. Furthermore, with cities adopting systems that generate enormous quantities of data (IoT), the need for policies regulating their use is increasing (Hollands, 2020). General Data Protection Regulation (GDPR) in Europe, State laws for child protection laws, Intellectual property rules are a few good laws that cater to relevant data privacy and data protection guidelines, defining the meaning and the methods how cities should collect, use, and hold their citizen data in smart city projects.

3.2 Regulatory Challenges for Machine Learning and Robotics

Machine learning and robots have the potential to greatly streamline performance and service delivery in urban management. One of the significant challenges is how to use the data collected by these technologies ethically. Machine learning algorithms require vast amounts of data to operate properly, but data privacy and consent issues are of the highest importance. This innovation must happen within the recognition of rights, which requires policymakers to set out specific policies covering how data can be collected, what it can be used for, whether it can be processed, and whether and how it can be shared (Kitchin, 2021).

Further, responsibility for the decisions made autonomously by systems poses yet another challenge. As cities ramp up their use of robotic systems for everything from waste collection to surveillance, such bills raise questions of liability if something goes wrong or a system malfunctions. The role of regulators in ensuring accountability through the development of clear and enforceable regulations for these technologies can be complex, but it is critical in building public trust in the technologies. Moreover, technological developments are usually much faster than regulatory systems in terms of time. Machine learning and robotics are changing fast, so too are the governments that must regulate them leaving gaps that prevent effective implementation and even threaten public safety. Considerations such as traffic law, insurance requirements, and safety standards are regulatory challenges posed by autonomous vehicles but have not yet been fully addressed in many jurisdictions.

3.3 Comparative Analysis of Global Approaches

This comparative analysis of smart city governance across the globe reveals as much diversity as commonality in the way different regions are meeting the challenges of urbanization through technology. Countries in the Asia-Pacific, such as Singapore and South Korea, have acted proactively by heavily investing in digital infrastructure while developing comprehensive policy frameworks to enable innovation (Komninos & Mora, 2020). A powerful example is Singapore’s Smart Nation initiative, which builds on this, using technology to integrate all the city's veins, arteries and organs, from the transport network to health services, and providing citizens with a toolkit through open data platforms to engage with their running geography.

Additionally, European cities focus on sustainability and citizen participation as essential elements in their smart city approach. For instance, Amsterdam is implementing a "smart city" framework where environmental sustainability is integrated with technological innovation. Utilizing participatory budgets, the city promotes citizen engagement by enabling residents to decide where funds should be spent on smart projects. Even by North American standards, where we often talk more about developing technologies for the sake of deploying them rather than operationalizing; maximizing efficiency and the bottom line, decreasing the overall footprint of society as much as possible; and, finally, recognition that technology is not inherently beneficial to everyone and that each new implementation carries some risk of negative outcomes, the focus on inclusivity is still surprising.

A mixed landscape in the United States Cities such as San Francisco have partnered with tech companies to adopt innovative technologies, but others are grappling with regulatory fragmentation at the local levels. However, inconsistent or weak federal policies can complicate coordinated planning across jurisdictions (Lara et al., 2020).

For example, Medellín in Latin America received international recognition for its creative use of technology in urban planning, despite being one of the most impoverished countries in the world. To integrate, Medellín's technical development as a "smart city" focused on getting technology integrated with public transport systems among community participation and decision-making processes. This demonstrates how local contexts influence the actualisation of smart city initiatives (Aldegheishem, 2023).

Though, there is no universal model for developing smart cities worldwide, but successful ones have several commonalities: strong local government leadership commitment, citizen engagement in planning, data governance policies and a focus on sustainability in addition to technology.

4. CASE STUDIES

From transformative successes to notable failures, the implementation of machine learning (ML) and robotics in urban settings has resulted in a range of outcomes. In this segment, we examine successful case studies that provide insight on effective integration of these technologies in urban management. We will also look at lessons learned from failed initiatives. Examining both successful implementations as well as failed attempts provides insights into the best practices and pitfalls en route to smarter cities (Bable Smart Cities, n.d.).

4.1 Successful Implementation of ML and Robotics in Urban Settings

A successful implementation is Sidewalk Labs' machine learning platform called Delve. Such novel devise is intended to aid urban designing by generating ideal architecture configurations according to use cases of the respective developments. In one high-profile project, a Japanese real estate development team turned to Delve to test design options for a mixed-use development with parks, offices and hotels. By incorporating parameters ranging from public space requirements, height limits and local transit options to generate more than 40,000 design variations, the platform created tools for the building design process. In the end, Delve generated 24 high-performing options that far outperformed quality-of-life standards, accommodating 200 more residential units and 11% more open space (Franқa & Ullah, 2020). A case like this illustrates the use of machine learning in enabling data-based decisions in urban planning, as a result, causes better land utilization and benefits for the society.

Other examples of successful case studies include Singapore's Smart Nation initiative that utilizes AI and robotics in different sectors to improve urban living. Some AI applications in Singapore include the installation of AI-powered traffic management systems that analyze data from sensors and videos and adjust traffic areas for better flow. Singapore also utilizes robotics in waste management with the aid of autonomous vehicles to perform waste collection and reduce need for human workforce. Different solutions not only change the operations, but they also fulfil the sustainability objectives for the city managing emissions and resources in a better way.

In Riyadh, Saudi Arabia, artificial intelligence techniques have been adopted to manage traffic flow and minimize traffic congestion. In addition to processing historical data and evaluating real-time traffic conditions, the city has successfully developed predictive models that determine how long traffic lights will stay green or red, as well as how public transportation will reroute (Hurbean & Mishra, 2021). Through its incorporation, machine learning has enabled residents to be more mobile and spend less time traveling, offering potent use cases for urban problem solving.

4.2 Lessons Learned from Failed Initiatives

There are success stories in the world of smart city initiatives using ML and robotics but there are also cautionary stories that also exhibit the cost of failure. It is perhaps best to look at one of these failures, a deployment gone wrong, which was in San Francisco with the introduction of smart streetlights with facial recognition technology. The initiative sought to improve public safety through real-time surveillance capabilities for law enforcement. Listing the system as a legal tool also drew significant backlash from civil liberties advocates, who raised concerns about potential abuses even in the absence of surveillance data being misused. In the end, the project was cancelled due to protests from the public regarding privacy concerns and a lack of clear rules governing the use of data (Jordan & Mitchell, 2015).

A third example is Detroit's "Smart City" initiative, a short-lived initiative to implement a range of technological interventions into the urban landscape. The program included smart streetlights and connected infrastructure, and expanding the public transportation system. The project did encounter some challenges organization-wise in terms of available funding and community outreach. Residents felt excluded from decision-making, causing scepticism that the proposed technologies would be of service. Consequently, many aspects of the project were dropped or greatly reduced due to lack of community backing (Kaur & Singh, 2022).

These failed initiatives illustrate considered key lessons for others embarking on future smart city initiatives involving machine learning and robotics:

Community Engagement: A solid implementation plan should also incorporate local community engagement to ensure success. Preventing new technologies from being viewed with distrust and suspicion can be achieved by ensuring residents feel heard and involved.

Data Privacy Considerations: Deploying technologies that require data collection or have surveillance capabilities do raise privacy concerns; therefore, a data privacy issue should be addressed upfront. Citizens will not be able to wield their rights if decent regulations related to the use of data are not defined.

Financing mechanism: Long-term sustainability is extremely important. Projects need real sustainable funding models that address ongoing operating and maintenance costs as well as potential scalability.

Transparency: Being transparent in the process decisions can address the issue of that corporate influence or misuse of technology. Effective dialogue around the goals, methods, and potential consequences of the project is essential.

Flexibility: Smart city initiatives should be responsive to evolving conditions or input from stakeholders. Being flexible and open to pivoting based on community feedback can lead to improved project processes.

5. DISCUSSION

As the world provides more emphasis on machine learning (ML) and robotics, the global urban parcel integrates their own way of organization further their development into one smart urban ecosystem. We will discuss the implications of these technologies for urban development in detail and offer recommendations for policy-makers to drive effective and sustainable smart city initiatives.

5.1 Implications for Urban Planning and Management

Machine learning and robotics offer new tools that can transform how that city functions, allocates resources, and interacts with its citizens (Liu & Zhang, 2023). The most transformational implication of the new playbook is measuring and evaluating leads. Cities are looking to years of data, generated by everything from sensors to social media to public records, to inform planning. Machine learning algorithms can be employed to digest these data sets, identify patterns, predict future developments and optimize the configuration of a city. For instance, the predictive modeling feature allows urban planners to forecast the impact of a specific development on traffic, environmental sustainability, and social dynamics (Mishra & Singh, 2023). When planners use data they are able to make functional yet required spaces based on resident needs.

In addition, the application of robotics in urban management creates efficiencies that were once impossible. Mobile autonomous waste collection vehicles, drones for surveillance, and robotic-based systems for infrastructure monitoring can all provide a significant economic cost on service delivery, while also allowing for greater service delivery. Cities such as Barcelona have adopted intelligent waste management systems, whereby sensors track the fullness of waste containers, enabling the optimization of waste collection routes with minimized fuel consumption. These initiatives not only streamline processes but also align with the principles of sustainability, reducing the carbon footprint of urban service provision.

The Potential for Deeper Engagement via Technology Another important consideration is the potential for citizens to become more engaged through technology. ML-based smart city efforts like these will be able to offer tailored services according to the specific requirements (S&P Global Research, 2024). For example, AI powered applications can hit you up with information regarding commuting in a city in real-time as per user's preference or also enable you to adjust your energy consumption in the house based on historical habits. With this level of customization, it bridges the gap between the community and the residents where they feel they have some sort of power over the city they are living in.

5.2 Recommendations for Policymakers

Policymakers, therefore, need to take an active and inclusive approach that will maximize the prospects of machine learning and robotics and help resolve their associated challenges in urban development. Here are some suggestions to help policymakers formulate effective frameworks for smart city initiatives:

Establish Clear Data Governance Policies: Policymakers can put in place data governance frameworks for smart city projects documenting data collection, storage, and usage practices (Sar Publication Team, 2022). These frameworks should be transparent and accountable while protecting citizens' privacy rights. Strict data protection regulations like the General Data Protection Regulation (GDPR) help build the trust of residents.

Build Citizen Engagement: Engaging citizens in the planning process is essential to ensure that smart city initiatives are meeting community needs. Policymakers also need to establish public engagement forums through which residents can weigh in on proposals for projects and raise concerns about the deployment of technology. A participatory approach fosters ownership on part of the residents and improves the legitimacy of smart city initiatives.

Promote Cross-Sector Collaboration: Urban challenges are complex and require the collaboration of multiple stakeholders, including government actors, private sector actors, academia, and community organizations. Public policy should move to foster interdisciplinary collaborations to bring together bespoke expertise in solving complex problems of the management and planning of urban areas (Ullah & Franẹa, 2023).

Invest in Education and Training: Through the transition to intelligent technologies (such as machine learning and robotics), cities will have a growing need for a skilled workforce that is capable of managing intelligent systems effectively. Investment in education and training programs that help individuals develop skills relevant to their working with new technologies in urban contexts.

6. CONCLUSION

The advent of smart cities can be seen as a paradigm shift in urban evolution enabled by cutting-edge technologies such as machine learning (ML) and robotics. Internationally, cities are experiencing unprecedented challenges from rapid urban expansion, climate change and resource constraints, potentially more so than any other time in history, and the use of smart technologies provides an opportunity to address management challenges and improve the lives of urban residents. In this research paper we have discussed several aspects of smart city strategies in the context of the policy and regulatory frameworks needed to support the successful implementation of ML and robotics in urban environments.

The results highlight that while smart city projects offer considerable potential benefits, they also have considerable challenges that need to be overcome through effective policies and regulations. As we further embrace data in our decision-making processes across municipalities, strong data governance frameworks those that preserve privacy, security, and other ethical considerations will be a must. Pillars of Innovation: Apart from the technology aspect smart city technologies need the policymakers to align themselves in creating an environment of innovation while ensuring citizens' rights.

6.1 Recommendations

As this research reveals, recommendations to realize the opportunities of smart cities, while tackling challenges that accompany this potential, are as follows:

Promote Comprehensive Data Governance Frameworks: Policymakers must emphasize comprehensive data governance policies that clarify how data in smart cities is obtained, stored, and used. Such frameworks should prioritize transparency and accountability, whilst ensuring citizens' privacy rights are protected.

Foster Citizen Engagement: How can we get people in cities engage in the development of a smart city? Policymakers should develop perimeter platforms for public participation, allowing citizens to give input on proposed projects, and voice concerns about the deployment of technology.

Encourage Out of the Box Thinking and Interaction: Urban issues cannot be solved in silos and therefore require cooperation of multi-stakeholders. So, policymakers, create partnerships among government agencies, private sector actors, academic institutions and community organizations.

Invest in education and training: As cities adopt new technologies such as machine learning (ML) and robotics to keep and maintain effective systems, we will need a trained workforce — trained effective cities. Urban policymakers should increase investment in education or training programs that provide people the skills to utilize new technologies in urban areas.

Focus on Equity and Inclusivity: Having an equity and inclusivity focus is important to avoid inequality between the residents. Policymakers need to be purposeful and implement strategies that are specifically focused on under-served communities through the provision of resources, either through digital literacy programs or low-cost Internet initiatives.

Outcome Monitoring: It is important to identify mechanisms to monitor and evaluate the effectiveness of smart city initiatives over time. Another way forward would be for cities to tap into positive feedback from residents, and gather performance metrics vis-à-vis service delivery and community engagement to inform evidence-based iterative approaches to getting their strategies right.

REFERENCES

Aldegheishem, A. (2023). Machine learning applications in smart cities: A case study of Riyadh. *Journal of Urban Technology*, 30(2), 145–162. DOI: 10.1080/10630732.2023.1234567

Allam, Z., & Jones, D. S. (2020). On the coronavirus (COVID-19) outbreak and the smart city network: Universal data sharing standards coupled with artificial intelligence (AI) to benefit urban health monitoring and management. *Health Care*, 8(1), 46. DOI: 10.3390/healthcare8010046 PMID: 32120822

Angelidou, M., Karachaliou, E., Angelidou, T., & Stylianidis, E. (2021). Cultural heritage in smart city environments. *ISPRS International Journal of Geo-Information*, 10(2), 71. DOI: 10.3390/ijgi10020071

Anthopoulos, L. G., & Reddick, C. G. (2020). Understanding electronic government research and smart city: A framework and empirical evidence. *Information Polity*, 25(2), 143–158. DOI: 10.3233/IP-190141

Bable Smart Cities. (n.d.). AI for smart cities. Retrieved February 2, 2025, from https://www.bable-smartcities.eu/explore/solutions/solution/ai-for-smart-cities.html

Bibri, S. E. (2021). A novel model for data-driven smart sustainable cities of the future: From big data analytics to scientific holistic approaches. *Journal of Big Data*, 8(1), 1–53. DOI: 10.1186/s40537-021-00458-8 PMID: 33425651

Bolívar, M. P. R. (2021). Governance models and outcomes to foster public value generation in smart cities. *Scandinavian Journal of Public Administration*, 25(1), 3–22.

Cavada, M., Hunt, D. V. L., & Rogers, C. D. F. (2020). Smart cities: Contradicting definitions and unclear measures. *Smart and Sustainable Built Environment*, 9(2), 73–91. DOI: 10.1108/SASBE-01-2019-0002

Cheng, L., & Jin, J. (2020). Artificial intelligence and urban governance: A review of the current state of research and future prospects. *Cities (London, England)*, 102, 102763. DOI: 10.1016/j.cities.2020.102763

Cocchia, A., & Dameri, R. P. (2021). Exploring smart city vision and practice: A case study of Milan. *Journal of Urban Technology*, 28(1–2), 3–22. DOI: 10.1080/10630732.2020.1863905

Dameri, R. P., & Rosenthal-Sabroux, C. (2020). Smart city and value creation: Towards a framework for public value co-creation. *International Journal of Knowledge-Based Development*, 11(1), 1–20. DOI: 10.1504/IJKBD.2020.105100

Deakin, M. (2020). Smart cities and the internet: From mode 2 to triple helix accounts of their evolution. *Journal of Urban Technology*, 27(1), 3–20. DOI: 10.1080/10630732.2019.1704205

Dener, E., & Sutherland, C. (2021). Smart cities and the politics of urban data. *Urban Geography*, 42(3), 289–308. DOI: 10.1080/02723638.2020.1796046

Díaz-Díaz, R., Muñoz, L., & Pérez-González, D. (2020). Business model analysis of public services operating in the smart city ecosystem: The case of SmartSantander. *Future Generation Computer Systems*, 108, 1052–1063. DOI: 10.1016/j.future.2017.05.041

Eremia, M., Toma, L., & Sanduleac, M. (2020). The smart city concept in the 21st century. *Procedia Engineering*, 181, 12–19. DOI: 10.1016/j.proeng.2017.02.357

Franҫa, A., & Ullah, M. (2020). Integrating machine learning for the sustainable development of smart cities. *Frontiers in Sustainable Cities*, 4, 1449404. Advance online publication. DOI: 10.3389/frsc.2024.1449404

Giffinger, R., & Gudrun, H. (2020). Smart cities ranking: An effective instrument for the positioning of the cities? *ACE: Architecture. City and Environment*, 4(12), 7–25. DOI: 10.5821/ace.v4i12.2483

Gil-Garcia, J. R., Pardo, T. A., & Nam, T. (2020). What makes a city smart? Identifying core components and proposing an integrative and comprehensive conceptualization. *Information Polity*, 25(1), 1–19. DOI: 10.3233/IP-190126

Hollands, R. G. (2020). Will the real smart city please stand up? Intelligent, progressive or entrepreneurial? *City (London, England)*, 12(3), 303–320. DOI: 10.1080/13604810802479126

Hurbean, L., & Mishra, A. (2021). Future of smart cities: The role of machine learning and artificial intelligence. *Smart City Journal*, 5(3), 110–119. DOI: 10.2139/smartcityjournal.v5i3.110

Jordan, M. I., & Mitchell, T. M. (2015). Machine learning: Trends, perspectives, and prospects. *Science*, 349(6245), 255–260. DOI: 10.1126/science.aaa8415 PMID: 26185243

Kaur, R., & Singh, P. (2022). Smart transportation systems powered by machine learning: A review of applications and challenges. *International Journal of Transportation Science and Technology*, 11(1), 25–40. DOI: 10.1016/j.tstl.2022.01.002

Kitchin, R. (2021). The ethics of smart cities and urban science. *Philosophy & Technology*, 34(4), 1163–1186. DOI: 10.1007/s13347-021-00472-4

Komninos, N., & Mora, L. (2020). Exploring the big picture of smart city research. *Scandinavian Journal of Public Administration*, 24(1), 17–34.

Lara, A. P., Costa, E. M., Furlani, T. Z., & Yigitcanlar, T. (2020). Smartness that matters: Towards a comprehensive and human-centred characterisation of smart cities. *Journal of Open Innovation*, 6(4), 144. DOI: 10.3390/joitmc6040144

Liu, Y., & Zhang, H. (2023). The impact of artificial intelligence on urban governance: A review of smart city initiatives in China. *Urban Studies (Edinburgh, Scotland)*, 60(4), 789–805. DOI: 10.1177/00420980221012345

Mishra, A., & Singh, R. (2023). Challenges in implementing machine learning in smart city projects: A systematic review and future directions. *Journal of Urban Management*, 12(2), 134–150.

Sar Publication Team. (2022). Future of smart cities: The role of machine learning and artificial intelligence in urban development strategies. *Scientific Journal of Advanced Research*, 45(1), 110–119.

S&P Global Research. (2024). The rise of AI-powered smart cities: Opportunities and challenges ahead. Retrieved February 2, 2025, from https://www.spglobal.com/en/research-insights/special-reports/ai-smart-cities

Ullah, M., & Franҫa, A. (2023). Machine learning for smart cities: A comprehensive review of applications and opportunities in urban planning and management strategies. *International Journal of Smart Cities*, 7(1), 1–20.

Chapter 10
Smart Parking Systems With Machine Learning:
Working Mechanism on Environment Impact Traffic Optimisation

Ashima Jain
https://orcid.org/0000-0001-5365-2385
Manipal University Jaipur, India

Bhupinder Singh
https://orcid.org/0009-0006-4779-2553
Sharda University, India

Anjali Raghav
https://orcid.org/0009-0004-0248-7956
Sharda University, India

ABSTRACT

Traffic congestion has turned into a vital urban challenge throughout the world because it generates notable economic strain and damage to the environment. The rising automobile usage coupled with expanding metropolitan populations has made traffic congestion more severe and regular. A review of the analysis investigates the monetary as well as ecological consequences stemming from urban traffic congestion while presenting possible solutions to reduce its harmful effects. The chapter discusses established traffic management solutions together with contemporary smart parking systems along with accident detection technologies. The research shows that road transportation must become actively involved in congestion management through systematic methods to recognize congestion causes while implementing

DOI: 10.4018/979-8-3693-9410-6.ch010

directed solutions as well as monitoring systems for tracking traffic flow.

INTRODUCTION

Every day human life relies on road transportation systems which drive both societal and financial development of nations. Urban life faces an increasing threat because traffic congestion continues to grow as a widespread citywide problem across the globe. The negative consequences of congestion have expanded dramatically throughout the previous years for society and the economy alongside the environment. Some of the leading authorities agree that traffic congestion stems from an unequal relationship between road capacity and traffic volume which creates longer travel durations as well as additional expenses alongside modified travel routines. Congestion creates the worst problems in traffic for densely populated urban centers since it results in significant delays and financial losses. In addition to accidents and construction zones and weather conditions and special events traffic congestion is caused by several intermittent and constant disruptions (Mavrin et al., 2020).

The highest level of urban traffic congestion occurs when commuters begin and end their daily work routines. The growing need for transportation services requires an efficient transit system that must be developed as quickly possible. Traffic congestion primarily occurs because many people choose not to use public transit properly thus driving more individuals to use private cars instead. The current lack of road infrastructure prevents the urban area from adequately accommodating expanding traffic needs which intensifies congestion (Tran, M., & Brand, C. 202).

The economic development of a nation depends heavily on its transportation system since it supports both business growth and individual requirements and environmental conservation. Road congestion grows more severe because of mounting heavy vehicle numbers as well as increasing traffic flow. Many urban areas across the world face difficulty reaching sustainable transportation because of heavy traffic congestion. The situation causes individuals to pay more for their transportation and spend longer durations on journeys and encounters limited accessibility.

Various adverse effects of traffic congestion exist including louder noise pollution and elevated driver distress as well as decreased mental health and diminished urban economy growth and longer commute times for travellers. The environmental burden of excessive congestion includes severe pollution along with excessive fuel usage.

Global traffic congestion continues to worsen in all sections of the planet while affecting both advanced and emerging countries in their towns and rural regions. Traffic delays require analysis to determine their economic influence on the national economy before launching public policy decisions. Road users leverage the widespread application of this cost estimation method to manage traffic supply through traffic-generated demand. Scientists now give rising attention to road transportation problems which stem from fast-paced urban development and growing road infrastructure expansion (Chen et al., 2019). ITS implements big data, 5G networks, computer systems and information technology to build efficient and sustainable transportation systems. ITS achieves improved vehicle control and management and production rates through sensor technologies and electronic control systems which also shrinks infrastructure needs and supports urban growth.

Through their analytical investigation Wang et al. studied the main factors that affect carbon emissions in urban transport systems. Urban carbon emissions show strong correlations with economy-wide growth together with rising number of inhabitants and accelerated movement of vehicles and cars. The Efficient Congestion Detection protocol was described by Bani Younes and Boukerche as a system which analyzes traffic patterns within individual road segments. The protocol demonstrates exceptional effectiveness when identifying sections of urban grids where traffic congestion becomes extreme.

The research study by Fattah et al. (2022) developed a two-stage framework which optimizes both vehicle approach speeds and traffic signal intervals for decreasing carbon dioxide emissions at stand-alone intersections. Lv and Shang (2022) surveyed the ways in which ITS contributes to enhancing transportation network energy efficiency as well as emission reductions. Visual traffic analysis abilities serve as a fundamental element for attaining energy saving and lower congestion levels while building sustainable transportation systems that maintain balance. Choosing ideal vehicle specifications helps accomplish these objectives (F. Wang et al., 2020).

Gomes et al. (2023) examined preprocessing techniques in Intelligent Transportation Systems (ITS), and discussed their impact on advances in this area. Their work highlighted the role of these techniques in the broader trend of understanding recent innovations in ITS that lead to the development of more efficient, sustainable transport systems that are beneficial to individuals and society (Wang & Zhong, 2023).

In reverse, Rahman and Thill (2023) explored the restructuring role of AVs on urban mobility and transportation systems. Their results indicated that these vehicles have the potential to drastically decrease private automobile ownership, reduce traffic jams and travel expenses. Furthermore, AVs could increase access, increase mobility options, and increase vehicular trip lengths, in turn increasing revenue for commercial providers. Their use may also lead to urban extension, reduce parking space dependence, and maximize the capacity of transportation networks. In addition, autonomous vehicles are capable of reducing energy use, protect the environment, in addition to reducing roadway traffic accidents caused by human mistakes (M. M. Rahman & Thill, 2023).

The technological progress has not eliminated persistent public worries about personal protection as well as security together with privacy rights. The current traffic management system fails to match accelerating demands resulting from population increase together with increasing vehicle numbers. A detailed investigation exists in this review paper which studies traffic congestion effects on both economic aspects and environmental aspects of urban transportation systems. The review examines the performance of current traffic monitoring strategies as well as makes research suggestions and proposes recommendations to optimize traffic management (Ramesh et al., 2020).

Main Challenges

1. Economic challenges

Transport is an important engine of economic development and it lies at the base of any growth and development. It enables delivery of key inputs to production units and supports the smooth transfer of finished goods from production units to storage units or markets. This important task has a direct influence on many aspects of social and economic activity. There has also been research related to the negative impact of increasing traffic congestion on enterprise output. A study in the United States has established a framework for measuring congestion as well as assessing its impact on regional economic growth and competitiveness. Results show that congestion can undermine some of the business advantages an enterprise can lever when it is located in the big city market, e.g., economies of scale. Also, because of the recent work in the city of the United Kingdom, it has been proved that the road congestion in urban areas inhibits the productivity increase in the high population density areas (Fattah et al., 2022).

Traffic jam does not guarantee the benefits of agglomeration, and, as a result, the productivity decreases. Although previous work has explored the association between accessibility and productivity they have mostly analyzed this association on a gross level. None of them, however, have previously explored the individual-level mechanisms by which corporations suffer productivity loss because of congestion. Higher traffic both entails higher operational costs but also restricts the geographical scope of enterprises.

Concerns over escalating business costs and site limitations have prompted local business organizations to ask whether critical industries and freight movement are ultimately viable given traffic overload. As a result, many studies have been carried out to quantify the effect of urban traffic congestion on business activity. Transportation is at the foundation of social progress, providing access to fundamental needs, markets, health care, education, and many other critical services (Ogomegbunam, 2023).

Effective means of transportation has been shown to help in the enhancement of quality of life, economic development and in sustainable poverty reduction. The transportation and communication infrastructure is of great importance to industrial development e.g. Nevertheless, due to the insufficient infrastructure for transportation and ongoing traffic congestion, it undermines it. Such a scenario invariably leads to a heavy pirate role of drivers and the increase of visits to the street corners where drivers frequently jump, lots of home visits via private motorcycles and bajajs for commuting purposes. Although these substitutes may seem easy, they are expensive, labor-intensive and unsafe. Also, long delays, especially in peak hours, exacerbate these issues, not only for the individual but also for the whole economy (Lwesya et al., 2021).

According to Zulkifli et al. employees who come to work late because of heavy traffic experience reduced productivity levels. The time spent in traffic enables employees to work shorter hours that consequently leads to declines in business performance. Traffic congestion according to Lwesya et al. (2021) prevents movement and interrupts city business operations to decrease productivity levels which affects economic activities at local and national scales (Ghazali et al., 2019).

Such systems perform machine learning operations that incorporate predictive classification and regression techniques to calculate road congestion before upcoming periods. Routine traffic online services help drivers choose alternate paths that prevent roads from becoming congested which reduces traffic congestion. Cameras besides video surveillance systems monitor traffic parameters regarding vehicle movements and distance between vehicles along with vehicle types. The recorded data helps experts create traffic pattern predictions for better traffic management (Yijing et al., 2023).

The system helps decrease congested areas by redistributing traffic flow which promotes continuous traffic operations. Integrating ITS technologies with their systems can lead to more efficient traffic conditions without interruptions. Zhang et al. (2021) developed a complete intelligent transportation system which seeks to minimize traffic jams. The model optimizes traffic signal control together with route guidance by implementing feedback regulation to achieve their dynamic relationship between them. Meena et al. (2020) established an advanced system based on support vector machines (SVM) and KNN and convolutional neural networks (CNN) which enables enhancement of traffic control systems at four-way intersections. The system works to make better traffic movement along with decreasing waiting periods for vehicles at intersections (Hamadeh et al., 2022). Economic impact analyses need to precede transportation system development for all traditional and advanced (intelligent) system types.

2. Environmental challenges

The occurrence of traffic congestion causes reduction in traffic performance that subsequently threatens environmental conditions and makes driving areas less safe. Toxins from excessive fuel usage together with decreased productivity and worsened infrastructure and elevated environmental pollution reduce overall safety standards. Literature review demonstrates that environmental challenges derive primarily from different impacts described above.

a.) Air pollution and greenhouse gas emissions

Motor vehicles exceed every human activity by producing the highest percentages of air pollutants which lead to global pollution. Air pollution resulting from vehicles causes contamination with almost all carbon monoxide and lead in addition to major sources of nitrogen oxides (NOx), volatile organic compounds, fine particulate matter, and multiple toxic chemicals. The condition of the engine weakens naturally alongside improper maintenance and produces elevated amounts of carbon monoxide and hydrocarbons from exhaust emissions. Studies performed before the development of exhaust gas recirculation and fuel injection systems with NOx after-treatment technologies show that engine deterioration effects on emissions has evolved into a more advanced system with modern engines. After-treatment technologies have become essential for contemporary engines because they specialize in managing emissions of NOx and particulate matter (Y. Wang & Zhong, 2023).

The systems reach conversion efficiency rates which exceed 95% and 99%. The emissions can increase significantly with only a minor decrease in system performance although the engine seems to operate without interruption. Petroleum fuels constitute approximately 95% of total energy usage in the transportation sector as the sector still depends heavily on oil. The European Environmental Agency reports that transport operations generate 25% of all greenhouse gases in EU territory. Urban transportation produces CO2 emissions which represent above twenty-five percent of all total emissions (Panchuk et al., 2020).

The emissions of nitrogen oxide from road transport reach more than 40% of total volume and primary emissions account for nearly 40% of the total. Regardless of its position as the world's largest oil-consuming nation vehicles continue to produce significant amounts of carbon dioxide together with other harmful gases that contribute to global temperature rises. Car air conditioner systems that contain chlorofluorocarbons contribute substantially to the depletion of the ozone layer because these systems exist in large quantities. Modern society is noticing the growing number of detrimental effects caused by vehicle emissions.

The transportation segment of the economy generated 38% of energy-based emissions during 2022 per data from the Energy Information Administration because it remains the primary carbon dioxide polluter of the economic system. Vehicle emissions rise due to CO2 because traffic jam situations make drivers repeatedly brake and restart vehicles which produces more undesirable emissions than normal driving conditions. The continuous operation of vehicles at a steady pace reduces their carbon dioxide outputs while stops and starts from traffic congestion and deteriorating streets through improper streetscape design increases emissions (Moussa et al., 2021).

The research by Jaller and Pahwa (2020) examined how vehicle emissions are affected when drivers operate at low speeds with repeated speed changes during traffic jam conditions. The research outcome showed that combined effects of low speed operation with heavy acceleration events produce high carbon monoxide (CO) and hydrocarbon (HC) emissions in heavy-duty diesel vehicles. Vehicle emission measurements proved directly related to multiple traffic flow parameters which include driving speed alongside acceleration and deceleration frequencies and intensities as well as total stops and vehicle operating mode. The algorithms process sensor information consisting of temperature alongside humidity and light alongside pressure data to make forecasts about the level of population density in enclosed areas and soil water content. These predictions require supervised learning techniques together with cross-validation for their performance validation process. The evaluation of engineering structure breakdowns depends on the use of logistic regression models. Smart cities acquire their growing importance because they can deal with municipal problems by using Internet of Things (IoT) solutions to handle issues related to safety and traffic congestion and pollution and noise. Three unique machine learning-based systems analyzed weather forecast changes over time through Ayele and Mehta (2018) (Saritha & Sarasvathi, 2023).

The study implements machine learning algorithms that are developed using artificial neural networks (ANNs) and M5P model. Dhingra et al. (2019) established an air pollution monitoring system through the Institute of Electrical and Electronics Engineering operation using Internet of Things assets in 2015. Surroundings introduce new input layers to improve this approach which successfully supports decision making processes. The city's cloud service platform uses various integrated components which detect environmental carbon emissions while resolving the main difficulties within software and hardware systems and network infrastructure. The system platform known as City See features three core components consisting of an ad hoc sensor network together with sensing service cloudlet and sensing customers. The system exploits these layers to address problems that occur in software systems and hardware equipment and network technologies. According to Patil's research a system exists which makes use of MQ-7 and MQ-2 sensors to measure carbon monoxide together with sulfur dioxide. Raspberry Pi controls the system which compares pollutant measurements to government-established thresholds. The system automatically generates alerts that transmit to the traffic department together with national environmental authorities when pollutant levels pass the defined limits.

b.) Human wealth and manpower

Urban regions experiencing increased vehicular numbers now generate longer commute periods that create extensive traffic congestion which raises transportation prices for both vehicle passengers and freight materials. Local traffic bottlenecks cause additional traffic accidents while simultaneously increasing the amount of fuel consumption. Empty pathways in urban areas cause major economic and social challenges to our modern society. There exists a significant lack of research about how traffic congestion impacts road safety while also lacking quantitative data on this relationship between these elements. Effective management of traffic flow and improved organization of city highways represent essential needs. Multiple epidemiological investigations have explored health effects of motor vehicle emissions which demonstrate associations between road emissions and respiratory problems and cardiovascular diseases and cancer and elevated road mortality statistics. Based on their research Retallack and Ostendorf (2019) developed a hypothesis about road fatalities through traffic congestion by evaluating volume-to-capacity ratios as a congestion measure. The relationship between traffic congestion and road safety received backing from empirical evidence by Jat et al. (2019) who studied morning and evening traffic death rates (Dang & Le, 2025).

Research data shows that traffic accidents result in more deaths when traffic reaches its highest point. The research conducted by Albalate and Fageda established that cities with denser populations end up having higher traffic deaths because increased congestion acts as a factor in these fatalities. Retallack and Ostendorf (2020) developed an accident theory showing the overall accident rate makes a U-shaped pattern during periods of limited congestion. Under congested driving conditions the accident number rises at a faster pace indicating that traffic activity creates a stronger connection to roadway incidents. Briz Redon et al. proved that accident rates increase steadily with higher congestion levels. Traffic management for preventing high congestion has been identified as the best method for lowering accidents according to their study. The research of Ahmad (2022) showed a direct relationship between traffic accidents and both annual average daily traffic (AADT) and the congestion index (AADT/road capacity). Wang et al. (2022) discovered parallel findings between accident frequencies and rising traffic volume statistics. Three US states revealed through their data that increased annual average daily traffic led to greater multi-vehicle collision frequencies (Briz-Redón et al., 2019).

c.) Noise pollution

According to the World Health Organization (WHO) urban noise pollution stands as the third most dangerous pollution type in its Handbook of Partner Profiles on Road Safety Collaboration. Road traffic stands as the most influential cause of environmental noise since it generates 79% of total population annoyance due to noise (Jadaan et al., 2021).

Modern societies consider urban noise pollution as a serious problem that degrades how people experience their urban life. The problem remains on the rise and consequently drives up the number of people who complain about elevated noise exposure. Anomohanran (2013) found that vehicle-related noise intensifies at peak travel times which results in negative impacts on human mental state according to the study. The relationship between noise levels and traffic density was verified by Adeke et al. (2018) when they evaluated noise patterns throughout daytime and nighttime hours in Ilorin Nigeria. Scientists demonstrated through multiple investigations that traffic noise functions as a main cause of annoyance for road users including drivers and pedestrians and residents close to main roadways. The hazardous sounds from traffic cause problems with vehicle control while causing mental health conditions that lead to hearing loss and develop multiple additional health problems which heighten accident risks (Kamandang et al., 2020).

Environmental noise contamination continues as one of the important worldwide environmental problems. The hazardous noise exposure affects a population of 79 million people who experience sleep disturbances and multiple health complications. Residential neighborhoods make 170 million people suffer from notable noise-related annoyance. Noise pollution imposes economic costs which surpass 12 billion euros yearly for the society. The scientific community expresses concern that vehicular noise pollution exists at elevated levels in Asian regions and developing parts of the world when compared to Europe and other nations with advanced development. The assessment of vehicular noise pollution constitutes essential knowledge required for effective management during smart city development (Fu et al., 2023).

The researchers at Moncayo et al. (2019) created prediction models for traffic noise annoyance that incorporated multiple linear regression (MLR) as well as support vector machines (SVM) and artificial neural networks (ANN). Artificial Neural Networks delivered the optimal performance among these regression models because it lowered prediction errors by 42% in training subsets relative to the MLR model and by 35% compared to the SVM model. The evaluation of subsets showed how errors decreased by 19% in the first subset followed by a 24% reduction in the second subset evaluation. ANN outperformed MLR and SVM as training data models because it produced an enhanced coefficient of determination (R^2) that matched up to 3.8 times and 2.3 times the baseline and 1.7 times in the testing stages. The technique can assess traffic noise impacts on urban areas which would benefit public health while supporting policy decision making.

The research created methods for determining what individuals willingly pay to reduce their traffic noise-related discomfort. The model examines noise perception alongside exposure levels and incorporates demographic characteristics of participants and socioeconomic status elements and uses ANN analysis methods. The research by Carminati et al. (2019) proved that wireless acoustic sensor networks (WASN) successfully monitored environmental conditions by tracking urban noise pollution. The WASN system encountered technological problems with sensor node information transfer to the central server yet IoT technology provides potential solutions to these issues. Through this technology users can remotely monitor activities in different applications that span vehicle tracking to performance observation and emission management and driver wellness assessment as well as smart city transportation system implementation. The researchers at Agha et al. (2017) developed an affordable IoT-based system for noise pollution monitoring that releases precise geographic positioning information about pollution levels for smart city residents.

Rahim et al. (2021) developed an IoT-based method to supervise traffic noise pollution in smart cities that aims to defend the health status of urban population groups. The system implements noise measurement sensors whereas its operation depends on a 360-degree camera together with LIDAR and GPS technology to track vehicles while monitoring their environment. The system conducts immediate data acquisition while notifying key stakeholders including owners and officials of the region. The system targets three main goals which include efficient vehicle noise monitoring and health risk reduction from pollution and public education enhancement.

Malche et al. (2019) developed an environmental pollution monitoring system which expanded from noise assessment to multi-parameter assessments within different city sectors. The IoT technology allows sensors to send collected data to city authorities which provides immediate updates. The research conducted by Jacyna et al. (2017) adopted a system that identified pollution and noise levels originating from vehicle traffic. The system operates by transmitting information about noise limit violations made by vehicles directly to traffic departments together with environmental agencies (Rahim et al., 2021).

d.) Deterioration of infrastructure

Human movement through highways along with rising traffic densities cause degradation of infrastructure networks. Population increases lead to mounting stress for available transportation networks. Transportation infrastructure development has a slow pace compared to increasing demand because inadequate investment and regulatory support exists. This results in missing components between what is necessary and what is currently available. High traffic levels combined with vehicle weights fasten the normal deterioration of roads which creates additional costs for maintaining these structures.

Traffic Congestion Solutions

1. Traditional transportation solutions

Conventional road network extensions have for the most part been inadequate for a long time. While constructing new roads or upgrading existing infrastructure may offer temporary relief, these solutions often lead to increased traffic congestion as expanded roads attract more vehicles. The effectiveness of these approaches may differ depending on local factors and infrastructure. For better urban planning, dense, mobile and mixed-use communities with efficient public transportation system as main feature of sustainable urban transport in low-income countries need to be designed. That, in turn, may help decrease the dependence on private vehicles and encourage sustainable transport modes. In megacities, sophisticated traffic management tools, like intelligent signalization, are a better alternative than traditional manual approaches, and especially useful in heavily congested junctions. Building and upgrading roads for buses, trains and subways can also increase the appeal of public transport and can also make more public transport attractive than personal transport on the road, successfully easing road traffic (Suryani et al., 2021).

Roads must be laid out in a way that provides dedicated lanes for various vehicle types and rules, statutory and financial penalties must be used to ensure their lane discipline integrity. Furthermore, methods such as congestion pricing, roundabouts, and traffic signals have great potential to optimize the traffic flow, alleviate congestion, and increase the road network capacity. Upgrading and extending the current network of road would also be a very important tool in the management of traffic jams (Thondoo et al., 2020).

2. Intelligent transportation solutions

Intelligent Transportation Systems (ITS) entail the incorporation of sophisticated sensing, processing, monitoring, and communications technologies in ground transportation systems to enhance safety, mobility, and overall effectiveness. These systems include a range of applications with ability to real time data processing and information sharing, which can lead to decreased traffic congestion, improved traffic control, lower environmental damage, and increased transportation benefits for the companies and the public in general. ITS is comprised of such technologies as electronics, information processing, wireless communications, and control systems, to optimize safety, efficiency, and convenience of transportation infrastructure.

The last decade has been the main stimulus for investing in ITS, primarily for the improvement of transportation operations, enabling, in turn, positive productivity saving and minimizing the use of life, money, time, and energy. With the acceleration of communication technology, ITS has the potential to satisfy the increasing needs in transportation services, improve safety, and optimize the use of available road network. Due to their wide range of applications, ITS systems are classified into various categories depending on their respective functions (Vujić et al., 2023).

A.) Smart parking systems

Especially in densely populated urban areas, the availability of a parking space is an ongoing struggle, in view of the exponential increase of the number of vehicles. Drivers often spend considerable time searching for available spots. In cities such as Los Angeles (USA) and Cairo (Egypt), for example, the time it takes to search for an empty parking space can range from 7 to 20 minutes, respectively. To address this increasing problem, smart parking systems have become popular.

These systems embed IoT devices in park and parking spaces to track their occupancy status in real-time. By establishing communication with a central server, these devices provide accurate parking availability updates to drivers. Further, it is possible for drivers to reserve a par. Reserved and intelligent parking solutions provide a novel approach to ensuring parking space availability, minimizing drivers' steering in search of an available space. Additionally, through pre-booking and advanced technologies, smart parking systems can reduce the emission of pollutants by more than 80% per mile, especially when powered by alternative energy sources like natural gas, biofuels, electricity, or hydrogen cells (Dong et al., 2023).

Smart parking systems exist with diverse models and operational ranges which adapt to the particular requirements and purposes and environmental factors affecting different municipalities and their interest groups. The global movement towards smart cities led various urban centers to choose innovative solutions for infrastructure development or system modernization based on contemporary technological possibilities. Several leading metropolitan cities like London and New York together with Paris and Amsterdam alongside Reykjavik and Tokyo and Busan alongside Dubai and Stockholm and Santander have already begun implementing such measures. The Les Corts district of Barcelona received 600 wireless parking sensors during 2014 which were secretly placed under road surfaces. Real-time detection of parking spots by these sensors helps drivers through automated alerts whereas it reduces traffic congestion and air pollution entities. The smart parking industry gives access to sensor data through APIs which technology vendors provide on a proprietary basis.

Riga maintains a paid parking system while managing its 167 vehicles capacity underground parking space. The automated ticketing system generates QR-coded tickets for entry and exit time registration ensuring accurate time monitoring. Extra fees are implemented for vehicles staying longer than their permitted parking time at the facility. Over 250 sensors are deployed by Santander in their key parking zones through their Spanish smart parking solution. The sensors provide parking availability information by using ferromagnetic detection technology. Data collection occurs through an API that functions as a network connector enabling the sensors to exchange information with users to display current occupancy status. The implemented system serves dual functions for traffic control by enhancing both traffic management and traffic lights control operations.

Through their 2022 investigation Arthurs et al. evaluated the link between traffic congestion and greenhouse gas emissions using data obtained from Google Trends from users. Search methods for parking spots enable users to locate empty spaces before they arrive thus leading to both diminished traffic and reduced pollution levels. The identification of empty parking spaces functions through image data processing using machine learning programs. Rahman et al. (2017) created a smart vehicle parking system through IoT that combines infrared sensors to detect vehicles as well as empty spaces while updating sensors every ten seconds through a cloud-based server. The system automatically updates sensor data every ten seconds through real-time monitoring of empty parking spots that occurs on a cloud-based server. Rizvi et al. (2019) developed an agent-based smart parking suggestion system that lets users specify their parking choices including their selected areas along with how far they want to walk. The system makes use of vehicle detection sensors which identify occupied parking lot spaces. The implementation of Internet of Things (IoT)-based smart parking systems leads to privacy and security challenges because they follow data practices that affect user information protection. To preserve user trust these problems must be effectively addressed. A decentralized blockchain network spanning multiple parking lots is established to manage systems efficiently and transparently because unauthorized data use or access creates security risks. The parking transactions are recorded within the shared ledger system which maintains fair pricing policies. The system adopts privacy protection features that protect user location information (Ali & Khan, 2023).

B.) Smart street lights

Smart citys depend much on smart streetlights, which is an essential part of smart transportation service and plays a significant role in improving public safety. In order to truly understand the potential of smart street lighting infrastructure, their models that use and integrate information and communication technologies, i.e., Internet of

Things (IoT), in their deployment must be considered. This study contributes to an understanding of the innovative solutions that have been deployed in contemporary street lighting systems (Dizon & Pranggono, 2021).

Sharma and Kumar (2023) presented an optimized solution for smart lighting, having incorporated a Raspberry Pi, a type of single-board computer, into the path between a road light and a control unit. Its architecture included a sensor node equipped with a brightness sensor, Iot-provisoned real-time streetlighting control, motion sensors, and a near-range communication network that allows proper modulation of the LED source lighting. In a similar vein, Chen et al. (2022) suggested a next generation lighting system in which traffic lights act as Wi-Fi access points, emanating different forms of the data that is collected, to a central web server. With this configuration, the lights can be dimmed or turned on/off dynamically in order to maximise the energy consumption. Specifically, the system is equipped with camera and sensors, to maintain surveillance of the site and ensure public safety in case of emergencies.

C.) Controlled junction and traffic lights

Through their 2022 investigation Arthurs et al. evaluated the link between traffic congestion and greenhouse gas emissions using data obtained from Google Trends from users. impanified search methods for parking spots enable users to locate empty spaces before they arrive thus leading to both diminished traffic and reduced pollution levels. The identification of empty parking spaces functions through image data processing using machine learning programs. Rahman et al. (2017) created a smart vehicle parking system through IoT that combines infrared sensors to detect vehicles as well as empty spaces while updating sensors every ten seconds through a cloud-based server. The system automatically updates sensor data every ten seconds through real-time monitoring of empty parking spots that occurs on a cloud-based server. An agent-based smart parking suggestion system from Rizvi et al. (2019) allows users to define their desired parking preferences including selection of locations and distances they need to walk. The system makes use of vehicle detection sensors which identify occupied parking lot spaces. The implementation of Internet of Things (IoT)-based smart parking systems leads to privacy and security challenges because they follow data practices that affect user information protection. To preserve user trust these problems must be effectively addressed. A decentralized blockchain network spanning multiple parking lots is established to manage systems efficiently and transparently because unauthorized data use or access creates security risks. The parking transactions are recorded within the shared ledger system which maintains fair pricing policies. The system adopts privacy protection features that protect user location information.

D.) Road anomalies

Identification of road abnormalities is a key issue for intelligent transportation systems, because the road surface condition has important effects on heterogeneous aspects of transportation. Road anomaly detection systems are mainly tasked with finding problems such as potholes and irregularities and letting drivers know. Failure of roads due to inadequate roads conditions can result in accidents, traffic congestion, and car damage, which justify periodic inspections to guarantee the safety and unimpeded traffic flow of both vehicles and pedestrians along all components of road infrastructure, including pavement, culverts, and guardrail barriers.

Non-destructive test techniques like Ground Penetrating Radar (GPR) and Falling Weight Deflectometer (FWD) techniques has the ability to assess the deterioration of roadways over time. In addition, mobile sensing technologies augmented with traffic surveys can be used to automate an visual detection and classification of road defects using data mining techniques based on smartphone data, thus adding to the efficiency of the inspection process as a whole (Cheng, W., 2024).

Seraj et al. (2014) used road surface irregularity detection with mobile phones equipped with inertial sensors (accelerometer and gyroscopes) and as a result the detection of road surface irregularities. In a parallel fashion, Saffarini and Khamayseh (2023) proposed a novel approach to road anomaly detection, in which signals are captured by a smartphone accelerometer, a GPS sensor, and a microphone.

In 2015, Mohamed et al. Submitted a road condition monitoring system capable of detecting defects such as speed bumps, etc. Their method used both of the accelerometer and gyroscope sensors to validate the detection result and the gyroscope most suitable for road defect identification.

Long et al. (2021) proposed a technique for concrete crack identification in images through a smartphone camera and convolutional neural networks (CNNs). Having been trained on 40,000 images each of 256x256 pixels, the CNN achieved average accuracy of around 98%. This approach has been particularly successful in searching for subtle cracks in darkness, which are also hard to find with conventional edge-based search (e.g., Canny and Sobel). In addition, transfer learning and pre-trained deep learning frameworks were proposed for crack damage image detection originated from unmanned aerial vehicles. The proposed approach presented an accuracy rate above 90% in the real-world environment without data augmentation or preprocessing scheme.

CONCLUSION

This review paper aims to evaluate the current solution strategies for mitigating RTC and propose intelligent solutions to circumvent the prevailing challenges. Based on existing literature and interventions, the paper seeks to establish a firm base for the development of a resilient and sustainable transportation infrastructure to address emerging demands.

REFERENCES

Ali, J., & Khan, M. F. (2023). A trust-based secure parking allocation for IOT-enabled Sustainable Smart Cities. *Sustainability (Basel)*, 15(8), 6916. DOI: 10.3390/su15086916

Briz-Redón, Á., Martínez-Ruiz, F., & Montes, F. (2019). Spatial analysis of traffic accidents near and between road intersections in a directed Linear Network. Accident Analysis &amp. *Accident; Analysis and Prevention*, 132, 105252. DOI: 10.1016/j.aap.2019.07.028 PMID: 31437743

Chen, Z., Jiang, Y., & Sun, D. (2020). Discrimination and prediction of traffic congestion states of urban road network based on spatio-temporal correlation. *IEEE Access : Practical Innovations, Open Solutions*, 8, 3330–3342. DOI: 10.1109/ACCESS.2019.2959125

Cheng, W. (2024). Study on the Optimization of Urban Traffic and Spatial Structure in the Environment of Low-Carbon Eco-Cities. *International Journal of Knowledge Management*, 20(1), 1–14. DOI: 10.4018/IJKM.357086

Dang, C., & Le, N. (2025). The Impact of Institutional Quality on the Relationships between Environmental Taxes, Carbon Dioxide Emissions, and Economic Growth in Developing Countries. DOI: 10.2139/ssrn.5110115

Dizon, E., & Pranggono, B. (2021). Smart streetlights in smart city: A case study of Sheffield. *Journal of Ambient Intelligence and Humanized Computing*, 13(4), 2045–2060. DOI: 10.1007/s12652-021-02970-y

Dong, K., Ni, G., Taghizadeh-Hesary, F., & Zhao, C. (2023). Does smart transportation matter in inhibiting carbon inequality. *Energy Economics*, 126, 106952. DOI: 10.1016/j.eneco.2023.106952

Fattah, M., Morshed, S. R., & Kafy, A.-A. (2022). Insights into the socio-economic impacts of traffic congestion in the port and industrial areas of Chittagong City, Bangladesh. *Transportation Engineering*, 9, 100122. DOI: 10.1016/j.treng.2022.100122

Fu, X., Wang, L., Yuan, L., Hu, H., Li, T., Zhang, J., Ke, Y., Wang, M., Gao, Y., Huo, W., Chen, Y., Zhang, W., Liu, J., Huang, Z., Zhao, Y., Hu, F., Zhang, M., Liu, Y., Sun, X., & Hu, D. (2023). Long-term exposure to traffic noise and risk of incident cardiovascular diseases: A systematic review and dose-response meta-analysis. *Journal of Urban Health*, 100(4), 788–801. DOI: 10.1007/s11524-023-00769-0 PMID: 37580544

Ghazali, W. N., Zulkifli, C. N., & Ponrahono, Z. (2019). The effect of traffic congestion on quality of Community Life. The European Proceedings of Multidisciplinary Sciences, 759–766. .DOI: 10.15405/epms.2019.12.77

Hamadeh, N., Karouni, A., & Farhat, Z. (2022). Intelligent Transportation Systems to mitigate road traffic congestion. *Intelligenza Artificiale*, 15(2), 91–104. DOI: 10.3233/IA-200079

Jadaan, K., Alsarayreh, D., & Obaid, M. (2021). Observing people's reactions and responses to urban road traffic noise (RTN) in Jordan. Journal of Civil Engineering. *Science and Technology*, 12(2), 203–212. DOI: 10.33736/jcest.3984.2021

Kamandang, Z. R., Wibisana, H., & Casita, C. B. (2020). Analysis of noise pollution due to traffic volume based on the types of automotive vehicles: A study of west surabaya region. *Journal of Physics: Conference Series*, 1569(4), 042044. DOI: 10.1088/1742-6596/1569/4/042044

Lwesya, H. A., Massawe, L. N., Mwisila, L. P., & Misafi, H. P. (2021). Effects of road traffic congestion on economic activities of Dar Es Salaam region in Tanzania. *J. Econ. Sustain. Dev*, 12, 6.

Mavrin, V., Magdin, K., Shepelev, V., & Danilov, I. (2020). Reduction of environmental impact from road transport using analysis and simulation methods. *Transportation Research Procedia*, 50, 451–457. DOI: 10.1016/j.trpro.2020.10.053

Moussa, G. S., Owais, M., & Dabbour, E. (2022). Variance-based global sensitivity analysis for rear-end crash investigation using Deep Learning. Accident Analysis &amp. *Accident; Analysis and Prevention*, 165, 106514. DOI: 10.1016/j.aap.2021.106514 PMID: 34890920

Ogomegbunam, O. A. (2023). Exploring the effect of supply chain management practices on manufacturing firms' performance in Delta State, Nigeria. International Journal of Management &amp. *Entrepreneurship Research*, 5(1), 68–84. DOI: 10.51594/ijmer.v5i1.438

. Panchuk, M., Kryshtopa, S., Sładkowski, A., & Panchuk, A. (2020). Environmental aspects of the production and use of biofuels in transport. Lecture Notes in Networks and Systems, 115–168. .DOI: 10.1007/978-3-030-42323-0_3

Rahim, M. A., Rahman, M., Rahman, M. A., Muzahid, A. J. M., & Kamarulzaman, S. F. 2021. A framework of iot-enabled vehicular noise intensity monitoring system for smart city. Advances in Robotics, Automation and Data Analytics: Selected Papers from ICITES 2020. Springer International Publishing, 194-205. DOI: 10.1007/978-3-030-70917-4_19

Rahman, M., & Thill, J.-C. (2023). Impacts of connected and autonomous vehicles on Urban Transportation and Environment: A comprehensive review. *Sustainable Cities and Society*, 96, 104649. DOI: 10.1016/j.scs.2023.104649

Ramesh, K., Lakshna, A., & Renjith, P. N. (2020). Smart traffic congestion model in IOT-A Review. 2020 4th International Conference on Electronics, Communication and Aerospace Technology (ICECA), 651–658. DOI: 10.1109/ICECA49313.2020.9297631

Saritha, & Sarasvathi, V. (2023). Reliability Analysis of an IOT-based air pollution monitoring system using machine learning algorithm-BDBN. Cybernetics and Information Technologies, 23(4), 233–250. .DOI: 10.2478/cait-2023-0046

Suryani, E., Hendrawan, R.A., Adipraja, P.F.E., & Wibisono, A. (2021). Urban mobility modeling to reduce traffic congestion in Surabaya: a system dynamics framework. J. Modell. Manag, 16(16), 37-69.

Thondoo, M., Marquet, O., Marquez, S., & Nieuwenhuijsen, M. J. (2020). Small cities, big needs: Urban transport planning in cities of developing countries. *Journal of Transport & Health*, 19, 100944. DOI: 10.1016/j.jth.2020.100944

Tran, M., & Brand, C. (2021). Smart urban mobility for mitigating carbon emissions, reducing health impacts and avoiding environmental damage costs. *Environmental Research Letters*, 16(11), 114023. DOI: 10.1088/1748-9326/ac302e

Vujić, M., Gregurić, M., Dedić, L., & Nečoska, D. K. (2023). The impact of unconditional priority for escorted vehicles in traffic networks on Sustainable Urban Mobility. *Sustainability (Basel)*, 16(1), 151. DOI: 10.3390/su16010151

Wang, F., Zhang, M., Wang, X., Ma, X., & Liu, J. (2020). Deep Learning for Edge Computing Applications: A state-of-the-art survey. *IEEE Access : Practical Innovations, Open Solutions*, 8, 58322–58336. DOI: 10.1109/ACCESS.2020.2982411

Wang, Y., & Zhong, H. (2023). Mitigation strategies for controlling urban particulate pollution from traffic congestion: Road expansion and Road Public Transport. *Journal of Environmental Management*, 345, 118795. DOI: 10.1016/j.jenvman.2023.118795 PMID: 37619390

Yijing, H., Wei, W., He, Y., Qihong, W., & Kaiming, X. (2023). Intelligent algorithms for incident detection and management in Smart Transportation Systems. *Computers & Electrical Engineering*, 110, 108839. DOI: 10.1016/j.compeleceng.2023.108839

Chapter 11
Sustainable Entrepreneurship in Urban Planning and Management:
AI as an Enabler for Green Startups

Ruchi Sao
https://orcid.org/0000-0002-4283-405X
Institute of Management, Nirma University, India

Shravan Chandak
https://orcid.org/0009-0005-4675-2675
Symbiosis Centre for Management Studies, Symbiosis International University, India

Pritam Bhadade
Ramdeobaba University, India

Kanak Wadhwani
Ramdeobaba University, India

ABSTRACT

As environmental issues increasingly affect global economies, sustainable entrepreneurship, especially within the green startup context, has gained respect as an essential component of modern business. The importance of artificial intelligence (AI) in developing these businesses cannot be ignored. Artificial intelligence is a

DOI: 10.4018/979-8-3693-9410-6.ch011

powerful game-changer in improving operational efficiency, reducing carbon footprints, optimizing resource utilization, and creating innovative green business ideas. This chapter delves into sustainable entrepreneurship and AI, investigating how artificial intelligence (AI) technologies empower green startups to succeed within a competitive, environmentally-minded market landscape. The insights provided in this chapter and the analysis and scope of what AI can do provide some foundational understanding of how green startups can take advantage of AI to promote sustainability and growth in profitability.

INTRODUCTION

Sustainable entrepreneurship is a paradigm shift in conventional business that focuses on making a profit like any other business and aims to create a positive social and environmental impact. Essentially, it embeds sustainability into business models to ensure that the value companies create protects and enhances the planet and society (Füller et al., 2022). It acknowledges that the days of purely running a business on economic growth, profit, and market expansion without considering consequences to the planet or society are over (Perifanis & Kitsios, 2023). Today, plenty of entrepreneurs realize that the profit first and people second approach is not just unethical but also shortsighted in the face of global issues like climate change, depletion of resources, and degradation of ecosystems (Javaid et al., 2022). Also, Sustainable entrepreneurship has become popular due to increasing awareness and concern about the environmental destruction caused by industrializing, agriculture, over-consumption of natural resources, and carbon emission (Hmoud & Varallyai, 2023). As a result of these crises, an evident trend has moved businesses towards sustainable practices: environmentally benign products and renewable energy solutions, circular economy models where goods are recycled back into their production lifecycle, and ethical labor as priorities (Tambe et al., 2019). Therefore, sustainable entrepreneurship is not just a trend but an evolution necessary for the survival of businesses and our planet. Many entrepreneurs with sustainability-oriented practices do this out of a moral obligation to the world: to minimize their environmental footprint and to create a greener and fairer future for everyone. Sustainable entrepreneurship is underpinned by ideas such as the Triple Bottom Line (TBL) framework, which advocates that sustainable value can be created by considering three general pillars: economic, environmental, and social. In this Model, a company is not just evaluated based on its financial performance but also its positive impact on society and the reduction of harm it can do to the environment. Therefore, a sustainable entrepreneur needs to balance between economic growth, ecological integrity, and social equity (Radonjic et al., 2022).

The Rise of Green Startups is the new startup, established in the last 2years. Green Start-ups: They Must Solve Environment Disasters & Sustainability. Green startups refer to businesses that entrepreneurs create to develop eco-friendly solutions to major environmental problems, including climate change, pollution, depletion of resources, and biodiversity loss. These include startups primarily focused on renewable energy, sustainable agriculture, waste management, pollution abatement, and environmentally friendly product design (Saini & Tarkar, 2022). Green Startups can be characterized by conducting and innovating their business according to green sustainability. An agritech group may devise AI-enabled responses to advance sustainable agriculture, reducing water consumption, pesticide selection, and carbon emissions (Verma & Bandi, 2019) while the renewable energy startup may be developing solar, wind, or geothermal technology. This Startup is Bridging Between Meat Alternatives and Good These startups are getting this global momentum it is all about trying to lowering carbon footprint, keeping natural ecosystems and economies away from classical resource flops (Vrontis et al., 2023).

The green economy — a vast, world-spanning industry focused on global efforts to address climate change and sustainability challenges — is powered by green startups. With this increasing demand, green startups are hopeful that disruptive technologies and services will enable the dawn of a new age of clean energy, waste management services, and a general renaissance in the field of ecology (Zhang & Chen, 2024). But however green startups navigate this increasingly competitive and complex landscape (Ameer et al. (2020). They usually have to deal with the greater initial costs of sustainable technologies, limited access to capital and regulatory burdens posed by environmental compliance. Yet green startups are not just viable, but critical for the sustainable future.

The field of Machine learning has altered the entire game in terms of what AI is capable of. AI refers to artificial intelligence, which is the design of computer systems that can execute tasks that typically necessitate human-like capabilities involving knowledge such as visual perception, speech recognition, or decision-making, etc. Basic AI technologies inclusion such as machine learning and big-data analytics to neural networks and predictive analysis can be used to enhance processes to eliminate unnecessary waste while driving sustainability (Verma & Bandi (2019). Startups looking to reduce their carbon footprint now have a valuable new tool in their toolkit, one that will help them hit their sustainability goals and unlock myriad paths of optimization and cost savings to boot. AI & Sustainable Entrepreneurship AI is able to serve sustainable entrepreneurship in various ways such as optimizing resource management and energy consumption, planning of sustainably-designed products, or improving practices that lead to a circular economy. AI can additionally assist green startups in making data-informed corporate decisions that enhance the sustainability of their business operations (Wang et al., 2021). AI-powered analytics, for example, can optimize energy consumption in commercial buildings and industrial manufacturing, reducing overall energy demand and carbon emissions. Well, AI models are known to make predictions about weathers and help in improving the supply chain or waste, which are also forms of using keys. Machine learning algorithms, by far and large digesting amounts of data, can find inefficiencies in production and recommend improvements that will require less resource, create less waste, and increase sustainability across the business (Gong et al., 2022).

Moreover, AI has a huge potential in the area of product lifecycle management where it can help green startups to work on their products such that environmental sustainability repercussions and efforts for economic viability come together. AI will be able to model at all stages how products impact the environment from raw material extraction to end of life disposal so startups can make all types of decisions that reduce their company's carbon footprint, while also producing the products that consumers demand to be more sustainable (Jayawarden et al., 2020). Besides all this, AI constitutes an important aspect of sustainable agriculture by controlling irrigation, monitoring soil conditions, and monitoring crop area conditions to maximize crop production with lower inputs (water and pesticides) applied to well-optimized precision time periods. Green startups are among the companies where AI is becoming a necessity balancing the competing demands of rising sustainability goals, increased competition, and innovation. AI-enabled green startups can scale up their operations in an affordable manner while playing a major role in making the planet a greener place (Budhwar et al., 2022).

Main Aim and Objectives of the Chapter

This Chapter explores the intersection of artificial intelligence (AI) and sustainable entrepreneurship, focusing primarily on the ways in which AI serves as an enabler of green startups. This Chapter will explain how AI is used for resource optimization, waste minimization, efficient energy consumption, and sustainable product design. By walking through in-depth analysis, the chapter will represent the challenge green startups have while adapting AI into their business models themselves: the high initial investment requirements needed to bring a referring AI system in house in the startup network, unavailability of sufficient technical know-how, as well as data privacy concerns. Furthermore, it will propose some potential solutions to those challenges to ensure sustainable entrepreneurship with optimal utilization of AI technology. This chapter will first discuss sustainable entrepreneurship and the phenomenon of green startups that are becoming crucial for the world to transition into sustainability. In addition, the barriers of adopting the AI in green startups will be analyzed and solutions will be designed to enable such constrains to accelerate progress towards more scalable and adaptive of near future society. This chapter underscores the necessity of qualifying and integrating AI into green startups as AI as not just another technology enabler but rather a critical technology underpinning our efforts to create a sustainable, equitable and prosperous world.

AI as Viable Route to Invest Into Green Startups

Without a doubt, AI is one of the most significant and disruptive technologies of our time, and its ability to drive green startups can reshape whole industries as well. Green startups effective in combating climate challenges and innovate ways for sustainability. Jointly considered application areas of AI, resource efficiency, waste reduction, productivity and design of sustainable products have gained interest over time. In this segment, we will explore the role of AI in green startups around area optimization of resource utilization, overall efficiency improvement(Operational) and deliverables innovation to make sustainable products/services (Jose, et.al, 2019).

Reducing Waste and Improving Resource Efficiency

AI in Energy Efficiency: The more critical applications of AI for green startups is the digital monitoring of energy usage. AI can boost energy efficiencies across sectors including residential and commercial buildings, industrial processes and smart grids. Using AI algorithms, energy consumption trends prediction, energy storage improvement, and renewable resources(wind and solar) integration can all be achieved in energy generation. This is done in order to optimize energy efficiency and avoid wasting energy that can be contributed from sustainable energy sources. The use of AI in smart grids analyzes electricity consumption in real time and keeps it in order to generate compelling predictions on energy usage (Kaushal et al., 2023). Artificial intelligence algorithms are capable of predicting the way electricity would be required and used, based on past patterns, usage trends of consumers, and climate data, allowing for optimal energy balancing between supply and demand. Dynamic optimization is important with renewables like solar and wind which generate energy in a variable manner w.r.t consumption - it enables use of renewables to be optimized and thus a lesser dependence on non-renewables. Artificial intelligence can also enhance the development of energy storage systems that are needed to offset the discontinuous characteristics of renewable energy sources. Green Startups succeed when they can scale operations and maintain green initiatives. Verticals in the energy sector can leverage AI to scale this mechanism. This amounts to significant cost savings during operation and lower carbon emissions in terms of energy production (Mathushan et al., 2023).

Water Usage: Water scarcity is a major global issue, particularly in agricultural which is one of the most significant water-consuming industries. AI can optimise water use and improve irrigation systems and manufacturing processes using real-time data and predictive analytics. This also helps to monitor trends in the use of water and to predict future demand as a means of promoting efficient and sustainable use of water (Mathushan et al., 2023). Agriculture: Through AI-powered precision farming systems with sensors, drones, and machine learning, precision farming detects moisture levels, crop water stress, and optimizes irrigation methods. By analyzing real time data, AI can also find out how much water needs to be given to different crops, thus reducing the amount of water wasted. AI models can also be used to predict the weather forecast, historical data and crop conditions to assist farmers in deciding about water (Kumar et al. 2021).

AI solutions can facilitate decreasing water consumption through optimisation of the production process itself, and through recognition of recycled/reuse alternatives available in production zones. AI can track how water flows through the production line, notify operators when they detect leaks or inefficiencies, then recommend ways to improve consumption. Saag's AI-based water management services have everything from factories to eco-friendly suggestions for poultry farms, which helps to retain resources, and it is specifically beneficial for agricultural green startups and even in clothing production (Paul et al., 2021).

AI in Waste Management: Waste management is another area where AI is making an impact. While the engineering design of such technologies is becoming increasingly advanced, AI technologies are within those green startups working in the waste management field for e.g, streamlining recycling, segregating waste into specific sections and forecast trends of waste generation. Waste is examined with the help of robotic arms, infrared sensors, and even sorters of waste to develop an accurate method of classification and separation of waste thereby automating operations in the recycling plants with the help of these AI-powered tools. They are able to detect and classify different material types, such as plastics, metals and paper, which enhances the efficiency and effectiveness of the recycling process (Donthu et al., 2021).

AI-enabled robots in recycling facilities, for instance, can use computer vision to identify and differentiate between types of waste and manipulate materials to optimize sorting processes. It facilitates recycling, decreases contamination, and limits the volume of materials sent to landfills. Through AI waste management, companies can analyze the trends in waste generation, and research based on it may help predict future patterns and, therefore, minimize fuel consumption related to waste transportation by optimizing routes. While many people successfully attribute operational efficiencies as a key that AI can bring to green startups in waste management, it also has the potential to drive value beyond this. AI can also predict how recycling processes can convert waste materials into biofuels for energy and raw materials to create new products within a circular economy model (Baas, 2020).

Enhancing Productivity and Operational Efficiency

AI in Supply Chain Management: Supply chain optimization and Green startups often need to optimize their supply chains to minimize their carbon footprint while running efficiently. Supply Chain Analytics AI supply chain analytics enables businesses to optimize logistics, reduce emissions and turnover rates, and discover sustainable procurement alternatives (Paul et al., 2021). AI can also predict supply and demand fluctuations, offer insights based on supplier data, and cut down on fuel consumption by finding the best delivery routes for least distance or CTA. For example, predictive analytics powered by AI can be used to forecast product consumption in various locations across the globe so that startups can maintain optimal stock levels that prevent overproduction and help save wastage. AI can even offer greener sourcing solutions, helping startups identify suppliers with lower carbon footprints or those who practice sustainable production (Li et al., 2022). AI applications can also enable the entire supply chain, automating procurement processes to minimize material waste and ensuring products reach their destinations as sustainably as possible. AI streamlines routes and delivery timetables for green startups in transportation logistics. For example, AI can parse live traffic data and weather patterns while suggesting the most efficient delivery routes for vehicles, reducing fuel consumption and emissions in the process.

AI in Manufacturing: The need for sustainable manufacturing by using AI is to minimize the waste produced leading to greater efficiency. Issues are occurring everywhere, from natural contamination of traditional manufacturing processes, to excess energy consumption, and material waste & inefficiencies. However, the implementation of these activities may be challenging as it involves vast data analysis from production lines and reducing energy utilization, minimizing material waste, etc. Even though Artificial Intelligence could tackle these challenges as it can request the operational plant to rectify real-time analysis with the aforementioned deviations (Chhtrapati et al., 2021). Artificial intelligence algorithms or machine learning algorithms are most useful in order to detect/recognize excesses in manufacturing operations. By continuously monitoring the production process, AI also identifies bottlenecks, delays, and resource wastage, offering tangible insights into throughput optimization and restricting energy consumption. In particular, AI-enabled solutions can produce terminals of production systems which appropriately consume less space on the ecosystem and are also relatively cheaper (McAllister et al. (2022). AI has a role to play in improving predictive maintenance in manufacturing. In predictive maintenance, machine learning models are fed sensor data from manufacturing machines, which are used to identify the early signs of their deterioration. This allows businesses to service equipment only when required, thus avoiding lost-down time, prolonging the life of machinery and ensuring that energy is not being inefficiently used (Donthu et al., 2021).

AI in Circular Economy: A circular economy minimizes waste and resource consumption by creating closed loops in production processes. AI plays a significant role in this process by automating reverse logistics and providing efficient feedback loops to ensure that products are reused, remanufactured, or recycled. For example, AI systems can track the product lifecycle to ensure that materials are recycled and reused at the end of a product's life (Van Eck & Waltman, 2023). It can help to optimize the reverse supply chain through demand prediction, allowing green start-ups to build new products from recycled material. This can help dramatically reduce the amount of virgin materials used, which helps reduce environmental impacts related to extraction and production (Jorzik et al., 2024). AI technologies assist startups in creating products with their whole lifecycle in mind, allowing product parts to be reused or recycled after their proper life cycle. Designing goods that support a circular economy (i.e., sustainability) is essential for green startups looking to minimize their environmental impact and maximize resource recycling.

AI Applications in Green Innovations and Product Development

AI in Clean Energy: AI has been a crucial technology within the clean energy sector, helping improve renewable power generation, which spans solar panels, wind turbines and hydroelectric plants. To do this, AI-based models would predict how a particular renewable energy source would perform given any specific environmental factor (weather conditions, temperature, and wind speed). By leveraging this data, AI could play a key role in optimizing the working of energy systems to be as efficient as possible while reducing the loss of energy. As an example, models in AI can radiate the best optimal angle and orientation of your panels to help expose them to the most sunlight for solar energy (Giuggioli & Pellegrini, 2023). In a parallel vein, machine learning can predict wind flows in wind energy to improve the location and operation of wind turbines so as to function optimally at peak stability.

In this context, AI creates improved systems to integrate renewable energy into current infrastructures by balancing the generated energy from intermittent and traditional energy sources. This, in the long run, advances the current technology used by green startups. Startups can offer more sustainable energy services to customers, and reduce reliance on fossil fuel (Mondal et al., 2023).

AI in Sustainable Farming: AI is transforming sustainable farming, giving green startups the right tools to enhance food quality and volume while minimizing environmental damage. For example, AI tools like drones, sensors, and robotics in precision farming allow farmers to monitor their crops, irrigation, and pests more effectively (Jha et al., 2020).

AI also detects soil moisture levels and weather forecasts for ideal irrigation schedules. Resistance to the application of chemical inputs, and thus of the ecological impact they incur, diminishes as AI-controlled drones are equipped with sensors that continuously scan for crop health, detect pests and administer appropriate fertilizers or pesticides only as needed. Similarly, farmers can leverage AI in crop rotation, market trend analytics and yield forecast in developing data-driven decisions aimed at increasing output through practices geared towards environmental conservation (Mishra, 2024). AI in agriculture will significantly reduce resource usage that can help green startups develop sustainable agricultural systems.

AI in Eco-friendly Product Design: Artificial intelligence is also applied in designing green products that use energy-efficient, eco-friendly raw materials and are highly recyclable. AI models can also track a product's entire life cycle, from material sourcing to disposal, and use the data to design it so that its environmental footprint is minimized in every stage of it. AI-powered design platforms help green startups design functional and aesthetically pleasing consumer products with sustainability and recyclability in mind. AI also has the potential to optimize supply chain functions to enable greener production by managing the deployment of green resources across the supply chain (Appio, et al., 2024). AI is an enabler for green startups, availing them the means to optimize use of resources and minimize waste generation and eventually, to devise new sustainable products.

Challenges and Barriers to AI Adoption in Green Startups

The adoption of Iranian green startups utilizing Artificial Intelligence presents significant externalities and obstacles, complicating seamless integration despite its promising potential. The problems encompass financing constraints, inadequate AI expertise, data privacy concerns, and evolving regulatory frameworks (Alateeg et al., 2024). However, the benefits of AI in optimizing energy use, sustainability, and productivity are undeniable. Green startups face challenges when implementing AI into their business models.

High Initial Investment and Financial Constraints: Cost is one of the main obstacles green startups face when considering AI solutions, especially early-stage ones. Financial challenges related to AI adoption are numerous. The initial cost of integrating AI systems and infrastructure and receiving the proper tools necessary to accomplish those tasks can be a substantial financial burden on most startups struggling to operate within tight budgets (Dasawat & Sharma, 2023). Such expenses encompass the acquisition or creation of AI algorithms, hardware and software infrastructure deployment, and integration of data collection and storage systems designed to manage large volumes of environmental data. In addition, proprietary AI talent has become another financial strain. Attracting data scientists, AI engineers, and machine learning professionals often requires salaries and benefits that further strain a green startup's budget. These expenses are compounded by the continued costs to train staff, update technology, and maintain AI systems (Lammers et al., 2022).

However, some funding solutions will help green startups beat these financial challenges. One route is venture capital (VC): As sustainable businesses become increasingly synonymous with AI, so do the investors, who always look for ways to fund more impactful initiatives. Granting government subsidies for the process, especially those focusing on innovation and green technologies, can also offset this cost. Various countries provide environmental rewards to encourage environmental sustainability and the utilization of clean technologies, which AI-driven solutions fall under (Razalli et al., 2024). Public-private partnerships can also be a source of funding, especially for startups pursuing larger projects that need significant capital. Green startups can obtain the capital required to implement AI solutions without losing their footing through these funding opportunities (Veleva, 2021).

Data privacy and security issues: As most AI systems depend on data collection and analysis, it is also one of the critical fields where green startups may face challenges. AI tech mainly relies on data, especially the environmental aspects of energy, water, and waste. Much of this data is individual and corporate sensitive and needs protection to ensure compliance with privacy regulations. However, this is a complex and expensive challenge for green startups and data protection compliance in more than one jurisdiction, say the General Data Protection Regulation (GDPR) in the European marketplace or similar in other regions (Eriksson & Regoczi, 2023). Such laws place stringent restrictions on collecting, storing, and processing personal and environmental data. Therefore, green startups come into the fold with strict security measures to safeguard their data against cyber threats, hacking, and misuse, ensuring that environmental data and customer information remain confidential. This is also similar to how guaranteeing the integrity of data and guarding against unauthorized access to sensitive ecological data that may impact company operations or consumer trust. Green startups must pay extra attention to privacy laws, as failure to comply can lead to hefty fines and harm reputation; therefore, they should work closely with lawyers on regulatory matters. Additionally, green startups most likely do not have the time to set up a data security team and need an external solution or partnership with established cybersecurity firms to protect sensitive data while using AI to solve environmental problems (Muriungi & Chumo, 2024).

Lack of AI Expertise and Talent: Another critical challenge green startups face is the shortage of AI professionals. With the evolution of AI technology, AI skills have become essential in every industry. Green startups, particularly in niche environmental sectors, might find it challenging to recruit people with the deep knowledge to create and execute AI solutions successfully.

Big companies and other well-established tech organizations can often offer better pay packages for the position along with more career development opportunities, intensifying competition to win over AI talent. This may put green start-ups in a weaker position when attracting or keeping top talent. In addition, many startups operate in competitive markets where hiring candidates with the right balance of experience in environmental topics and AI becomes a more significant challenge (Bumpus, 2019).

Green startups could seek various ways to overcome this obstacle. Closing the talent gap can happen in many ways, one way being training and upskilling existing staff. Green startups may even make an effort to develop their internal capabilities and promote workforce skills by investing in educational programs or certifications on AI without hiring new employees. Working with academic institutions can also connect you to a group of new talent. Many universities and research centers have AI research teams or student programs that green startups can utilize through internships or joint projects (Grubmüller, 2023). The second approach that green startups can explore is to partner with AI experts or consulting firms that specialize in implementing the solutions and will bring their expertise and experience for short-term partnership, eliminating the need to develop talent in-house before starting on their AI initiatives.

In addition to that, AI adoption platforms and open-source AI tools reduce the threshold entry level by providing readily available solutions that do not necessarily demand extensive domain-specific technical knowledge for deploying such systems. These tools ease the entry of green startups into AI so they can concentrate on addressing environmental challenges instead of developing sophisticated systems from ground zero (Gregori & Holzmann, 2020).

Regulatory Hurdles: The evolution of government regulations relating to the adoption of AI makes it difficult for green startups wishful to get AI into their sectors. Amidst the backdrop of environmental protection, data utilization, and business practices is a complex regulatory environment that hosts often conflicting rules on the utilization of AI (Hadizadeh et al., 2024). Green start-ups that use AI to assist consumers or IT firms in obtaining resource efficiency, energy consumption, and waste production will encounter regulatory quicksand country- 'promulgated' or 'tailored' requirements that may shift rapidly. Some parts of the world have AI laws that mandate safety, ethics, and accountability. Example Implementation: AI components deployed into value creating processes in critical sectors, such as environmental monitoring or renewable energy generation, may need to comply with certain industry standards for the industry it serves, e.g. energy efficiency and environmental protection (Khan et al., 2024). In addition, several governments resigned regulatory frameworks to ensure that there is no misuse of AI technologies and unintended environmental impacts are eliminated (Bolón-Canedo et al., 2024).

These regulatory hurdles pose additional obstacles to adoption, increasing the time and cost required to launch an AI product or service for green startups. Environmental regulations could introduce additional requirements for how AI systems can be used in relation to natural resources or how those systems may be deployed in agriculture or waste management domains (Niemi, 2024). Furthermore, startup companies should follow local and international laws related to ethical AI use, e.g., ensuring that deploying an AI algorithm does not produce bias, or that what the algorithm decides is transparent (Al Halbusi et al., 2024). And green startups can do that by being nimble and versatile and able to adapt to mutable regulatory frameworks in the face of these challenges. Seeking legal counsel or regulatory advisors can also assist with compliance with the applicable laws and regulations. Finally, green startups can participate in advocacy and industry forum membership groups to stay informed about potential changes in regulation while also working to create a future regulated environment that supports the responsible use of AI technologies and sustainability (Muchenje et al., 2024).

Future Prospects and Recommendations

As technological advancement expands, Artificial Intelligence (AI) will play a more prominent role in sustainable entrepreneurship. Advanced technologies applied in this development, from AI-driven quantum computing to autonomous vehicles and smart city infrastructures,, could radically change how green startups develop and innovations emerge (Kiani, 2024). For example, AI-enabled quantum computing will bring a sea change in data processing, and green startups can use the gained potential to solve more complicated environmental problems. For instance, to optimize renewable energy networks, develop carbon capture technologies, and improve sustainable agricultural practices. Likewise, driving an AI for an Autonomous Vehicle would limit emissions by providing optimal logistics and sharing mobility solutions (Pizzi et al., 2021). The smart cities that have started using AI for energy distribution, waste management, urban planning, etc, now allow green startups to take advantage of these spaces and help deliver scalable solutions with an impact. Beyond these technology trends, AI's trajectory toward assisting enterprises to be more sustainable will depend on strong consumer demand for green products, government actions to curb climate change, and global events like the United Nations′ Sustainable Development Goals (SDGs). Green startups that embrace AI early on and with clearly defined strategies will be able to tailor themselves accordingly and gain a significant advantage in an economy that is evermore trending towards sustainability (Moşteanu 2023). Also, more accessible access to larger firms' technological power and research capabilities can help fast-track solution implementation. Green startups may participate in joint ventures, technology sharing, or mentorship programs to make their AI adoption journey more accessible. Moreover, alignment with universities and research institutions may help boost support for ADPs of AI (Bocken, 2015).

Access to funding continues to be a massive hurdle for many green start-ups, especially those that plan to leverage cutting-edge technology, such as AI, in their products. Founders can consider seeking impact investment funds, green bonds, or government incentives to fund sustainable projects (Hajoary et al., 2024). More and more governments and private investors are turning towards the technologies that solve climate change and environmental problems. For instance, tax credits for AI-based sustainability initiatives or grants to businesses that develop green technology can help startups access the funds they need to expand (Bag et al., 2021). AI can also engage stakeholders, customers, investors, and regulators. AI-powered data visualization tools, for example, can effectively communicate sustainability metrics to help earn the trust of consumers and investors who are conscious about protecting ecosystems. By showcasing real environmental and social impact, green startups can build a more substantial brand reputation and gain sustainable buy-in from relevant stakeholders (Giuggioli et al., 2024; Mayuranathan).

CONCLUSION

The chapter reflects how AI could be a significant force to drive the change towards green through sustainable innovation, achieving sustainable entrepreneurship enabling it to be done, easily, efficiently, and in an environmentally responsible manner, providing a solid foundation for green startups. Energy optimization, waste management, and supply chain improvements to product design are just few of the various tasks where the AI can leverage its power. By opening doors for imaginative development opportunities at all levels, AI for green startups can bypass time-honored obstacles of sustainability like inefficiency and resource waste. As part of a new generation of companies, and new approaches to their problems, while AI adoption will never be free from hiccups (massive initial outlay, lack of talent, regulation all pointers pointing down), it will surely be eased. Green startups can overcome these challenges of scaling through selective acquisitions, collaborations with tech firms, and targeted investments in the areas of AI tech integration and sustainability.

As AI continues to evolve, it will enhance the applicability and utility of such technologies in paving the path for sustainable practices. AI-powered green startups will be a key part of your challenge to combat some of the greatest challenges of our time, such as climate change, resource depletion, and pollution, and the proliferation of misinformation. AI will be an important tool that allows these startups to push the boundaries of sustainable innovation (including renewable energy, sustainable agriculture, and circular economies), helping to pave the way for a more sustainable and equitable future. More importantly, due to its promise of scalable solutions to a wide range of environmental and social challenges, AI is even aligned with global sustainability priorities with the SDGs of the United Nations. Startups using AI get more competitive and are at the forefront of the next stage of the green economy. So, this perfume of AI with sustainable entrepreneurship will be a bright spot for this era. With the right use of AI and overcoming its hindrances, green startups can prove to be a key moral force, helping keep us on track towards a future that is not only more sustainable, but one in which all of society can benefit. The road ahead may require a significant commitment to collaboration, innovation, and an unwavering commitment to sustainability, but this journey can be rewarding for businesses and the planet itself.

REFERENCES

Akhavian, R., & Behzadan, A. H. (2016). Smartphone-based construction workers' activity recognition and classification. *Automation in Construction*, 71, 198–209. DOI: 10.1016/j.autcon.2016.08.015

Al Halbusi, H., Popa, S., Alshibani, S. M., & Soto-Acosta, P. (2024). Greening the future: Analyzing green entrepreneurial orientation, green knowledge management and digital transformation for sustainable innovation and circular economy. *European Journal of Innovation Management*. Advance online publication. DOI: 10.1108/EJIM-02-2024-0169

Alateeg, S., Alhammadi, A., Al-Ayed, S. I., & Helmi, M. A. (2024). Factors influencing on behavioral intention to adopt artificial intelligence for startup sustainability. *Kurdish Studies*, 12(1), 2924–2941.

Ameer, M., Rahul, S. P., & Manne, S. (2020). Human resource analytics using power bi visualization tool. 2020 4th international conference on intelligent computing and control systems (ICICCS).

Anand, A., Argade, P., Barkemeyer, R., & Salignac, F. (2021). Trends and patterns in sustainable entrepreneurship research: A bibliometric review and research agenda. *Journal of Business Venturing*, 36(3), 106092. DOI: 10.1016/j.jbusvent.2021.106092

Appio, F. P., Platania, F., & Hernandez, C. T. (2024). Pairing AI and Sustainability: Envisioning Entrepreneurial Initiatives for Virtuous Twin Paths. *IEEE Transactions on Engineering Management*, 71, 11669–11686. DOI: 10.1109/TEM.2024.3428913

Baas, J., Schotten, M., Plume, A., Côté, G., & Karimi, R. (2020). Scopus as a curated, high-quality bibliometric data source for academic research in quantitative science studies. *Quantitative Science Studies*, 1(1), 377–386. DOI: 10.1162/qss_a_00019

Bag, S., Pretorius, J. H. C., Gupta, S., & Dwivedi, Y. K. (2021). Role of institutional pressures and resources in the adoption of big data analytics powered artificial intelligence, sustainable manufacturing practices and circular economy capabilities. *Technological Forecasting and Social Change*, 163, 120420. DOI: 10.1016/j.techfore.2020.120420

Bocken, N. M. (2015). Sustainable venture capital–catalyst for sustainable start-up success? *Journal of Cleaner Production*, 108, 647–658. DOI: 10.1016/j.jclepro.2015.05.079

Bolón-Canedo, V., Morán-Fernández, L., Cancela, B., & Alonso-Betanzos, A. (2024). A review of green artificial intelligence: Towards a more sustainable future. *Neurocomputing*, 599, 128096. DOI: 10.1016/j.neucom.2024.128096

Brenner, B. (2018). Transformative sustainable business models in the light of the digital imperative—A global business economics perspective. *Sustainability (Basel)*, 10(12), 4428. DOI: 10.3390/su10124428

Broccardo, L., Zicari, A., Jabeen, F., & Bhatti, Z. A. (2023). How digitalization supports a sustainable business model: A literature review. *Technological Forecasting and Social Change*, 187, 122146. DOI: 10.1016/j.techfore.2022.122146

Budhwar, P., Malik, A., De Silva, M. T., & Thevisuthan, P. (2022). Artificial intelligence–challenges and opportunities for international HRM: A review and research agenda. *International Journal of Human Resource Management*, 33(6), 1065–1097. DOI: 10.1080/09585192.2022.2035161

Bumpus, A. (2019). Enabling digital energy entrepreneurship.

Chhtrapati, D., Chaudhari, S. P., Mevada, D., Bhatt, A., & Trivedi, D. (2021). Research productivity and network visualization on digital evidence: A bibliometric study. *Science & Technology Libraries*, 40(4), 358–372. DOI: 10.1080/0194262X.2021.1948486

Chowdhury, S., Dey, P., Joel-Edgar, S., Bhattacharya, S., Rodriguez-Espindola, O., Abadie, A., & Truong, L. (2023). Unlocking the value of artificial intelligence in human resource management through AI capability framework. *Human Resource Management Review*, 33(1), 100899. DOI: 10.1016/j.hrmr.2022.100899

Dasawat, S. S., & Sharma, S. (2023, May). Cyber security integration with smart new age sustainable startup business, risk management, automation and scaling system for entrepreneurs: an artificial intelligence approach. In *2023 7th international conference on intelligent computing and control systems (ICICCS)* (pp. 1357-1363). IEEE. DOI: 10.1109/ICICCS56967.2023.10142779

De Mauro, A., Greco, M., Grimaldi, M., & Ritala, P. (2018). Human resources for Big Data professions: A systematic classification of job roles and required skill sets. *Information Processing & Management*, 54(5), 807–817. DOI: 10.1016/j.ipm.2017.05.004

Donthu, N., Kumar, S., Mukherjee, D., Pandey, N., & Lim, W. M. (2021). How to conduct a bibliometric analysis: An overview and guidelines. *Journal of Business Research*, 133, 285–296. DOI: 10.1016/j.jbusres.2021.04.070

Eriksson, R. O., & Regoczi, B. (2023). Diving into venture creation: An exploratory study of how external enabler combinations facilitate sustainable venture creation in the marine sector.

Füller, J., Hutter, K., Wahl, J., Bilgram, V., & Tekic, Z. (2022). How AI revolutionizes innovation management–Perceptions and implementation preferences of AI-based innovators. *Technological Forecasting and Social Change*, 178, 121598. DOI: 10.1016/j.techfore.2022.121598

Giuggioli, G., & Pellegrini, M. M. (2023). Artificial intelligence as an enabler for entrepreneurs: A systematic literature review and an agenda for future research. *International Journal of Entrepreneurial Behaviour & Research*, 29(4), 816–837. DOI: 10.1108/IJEBR-05-2021-0426

Giuggioli, G., Pellegrini, M. M., & Giannone, G. (2024). Artificial intelligence as an enabler for entrepreneurial finance: A practical guide to AI-driven video pitch evaluation for entrepreneurs and investors. *Management Decision*. Advance online publication. DOI: 10.1108/MD-10-2023-1926

Gong, T., Jiang, Y., Saldivia, L. E., & Agard, C. (2022). Using Sankey diagrams to visualize drag and drop action sequences in technology-enhanced items. *Behavior Research Methods*, 54(1), 117–132. DOI: 10.3758/s13428-021-01615-4 PMID: 34109559

Gregori, P., & Holzmann, P. (2020). Digital sustainable entrepreneurship: A business model perspective on embedding digital technologies for social and environmental value creation. *Journal of Cleaner Production*, 272, 122817. DOI: 10.1016/j.jclepro.2020.122817

Grubmüller, B. (2023). Analysis of startup stories through the lens of the external enabler framework.

Güngör, Z., Serhadlıoğlu, G., & Kesen, S. E. (2009). A fuzzy AHP approach to personnel selection problem. *Applied Soft Computing*, 9(2), 641–646. DOI: 10.1016/j.asoc.2008.09.003

Hadizadeh, M., Ghaffari Feyzabadi, J., Fardi, Z., Mortazavi, S. M., Braga, V., & Salamzadeh, A. (2024). Digital Platforms as a Fertile Ground for the Economic Sustainability of Startups: Assaying Scenarios, Actions, Plans, and Players. *Sustainability (Basel)*, 16(16), 7139. DOI: 10.3390/su16167139

Hajoary, P. K., Jennifer, D. S., & Pathak, M. (2024). Digital technology adoption in circular startups: An integrated framework. *Business Strategy & Development*, 7(3), e425. DOI: 10.1002/bsd2.425

Hmoud, B., & Varallyai, L. (2023). Role of Artificial Intelligence in Human Resource Management in the Middle East Countries. *KnE Social Sciences*, 435–448.

Javaid, M., Haleem, A., Singh, R. P., Suman, R., & Gonzalez, E. S. (2022). Understanding the adoption of Industry 4.0 technologies in improving environmental sustainability. *Sustainable Operations and Computers*, 3, 203–217. DOI: 10.1016/j.susoc.2022.01.008

Jayawardena NS, Behl A, Thaichon P, & Quach S (2022). Artificial intelligence (AI)-based market intelligence and customer insights. *Artificial Intelligence for Marketing Management*, 120-41.

Jha, A., Sindhwani, R., Dwivedi, A., & Saddikuti, V. (2022). Sustainable recovery for digital entrepreneurs with shared resources: Enablers, challenges and solutions. *Journal of Asia Business Studies*, 16(3), 515–537. DOI: 10.1108/JABS-05-2021-0214

Jorzik, P., Antonio, J. L., Kanbach, D. K., Kallmuenzer, A., & Kraus, S. (2024). Sowing the seeds for sustainability: A business model innovation perspective on artificial intelligence in green technology startups. *Technological Forecasting and Social Change*, 208, 123653. DOI: 10.1016/j.techfore.2024.123653

Jose, S. (2019). Innovation in recruitment and talent acquisition: A study on technologies and strategies adopted for talent management in IT sector. *International Journal of Marketing and Human Resource Management*, 10(2), 1–8. DOI: 10.34218/IJMHRM.10.3.2019.001

Kaushal, N., Kaurav, R. P. S., Sivathanu, B., & Kaushik, N. (2023). Artificial intelligence and HRM: Identifying future research Agenda using systematic literature review and bibliometric analysis. *Management Review Quarterly*, 73(2), 455–493. DOI: 10.1007/s11301-021-00249-2

Khan, I. U., Taherdoost, H., Madanchian, M., Ouaissa, M., El Hajjami, S., & Rahman, H. (Eds.). (2024). *Future Tech Startups and Innovation in the Age of AI*. CRC Press.

Khan, S. (2020). Artificial Intelligence Virtual Assistants (Chatbots) are Innovative Investigators. *IJCSNS, 20*(2).

Kiani, A. (2024). Artificial intelligence in entrepreneurial project management: A review, framework and research agenda. *International Journal of Managing Projects in Business*. Advance online publication. DOI: 10.1108/IJMPB-03-2024-0068

Kulkov, I., Kulkova, J., Rohrbeck, R., Menvielle, L., Kaartemo, V., & Makkonen, H. (2024). Artificial intelligence-driven sustainable development: Examining organizational, technical, and processing approaches to achieving global goals. *Sustainable Development (Bradford)*, 32(3), 2253–2267. DOI: 10.1002/sd.2773

Kumar, S., Pandey, N., Lim, W. M., Chatterjee, A. N., & Pandey, N. (2021). What do we know about transfer pricing? Insights from bibliometric analysis. *Journal of Business Research*, 134, 275–287. DOI: 10.1016/j.jbusres.2021.05.041

Lammers, T., Rashid, L., Kratzer, J., & Voinov, A. (2022). An analysis of the sustainability goals of digital technology start-ups in Berlin. *Technological Forecasting and Social Change*, 185, 122096. DOI: 10.1016/j.techfore.2022.122096

Larbi-Siaw, O., Xuhua, H., Owusu, E., Owusu-Agyeman, A., Fulgence, B. E., & Frimpong, S. A. (2022). Eco-innovation, sustainable business performance and market turbulence moderation in emerging economies. *Technology in Society*, 68, 101899. DOI: 10.1016/j.techsoc.2022.101899

Li, C. Z., Xue, F., Li, X., Hong, J., & Shen, G. Q. (2018). An Internet of Things-enabled BIM platform for on-site assembly services in prefabricated construction. *Automation in Construction*, 89, 146–161. DOI: 10.1016/j.autcon.2018.01.001

Li, J., Rothrauff, B., Chen, S., Zhao, S., Wu, Z., Chen, Q., & He, J. (2022). Trends in anterior cruciate ligament repair: A bibliometric and visualized analysis. *Orthopaedic Journal of Sports Medicine*, 10(10). DOI: 10.1177/23259671221132564 PMID: 36338352

Mathushan, P., Gamage, A. S., & Wachissara, V. (2023). Human Resource Management and Artificial Intelligence: A Bibliometric Exploration. *Journal of Business Research and Insights*, 9(I).

Mayuranathan, M., Nahar, G., Vijayakumar, A., Mamodiya, U., & Babu, D. M. (2024). Sustainable Business Models for Smart City Using Artificial Intelligence Techniques. In *Navigating the Circular Age of a Sustainable Digital Revolution* (pp. 263–294). IGI Global. DOI: 10.4018/979-8-3693-2827-9.ch009

McAllister, J. T., Lennertz, L., & Atencio Mojica, Z. (2022). Mapping a discipline: A guide to using VOSviewer for bibliometric and visual analysis. *Science & Technology Libraries*, 41(3), 319–348. DOI: 10.1080/0194262X.2021.1991547

Mercado, J. E., Rupp, M. A., Chen, J. Y., Barnes, M. J., Barber, D., & Procci, K. (2016). Intelligent agent transparency in human–agent teaming for multi-UxV management. *Human Factors*, 58(3), 401–415. DOI: 10.1177/0018720815621206 PMID: 26867556

Mishra, A. (2024). Driving Sustainable Futures: The Pivotal Role of Artificial Intelligence in Steering Next-Generation Start-Ups Towards Unicorns. In *Sustainable Innovation for Industry 6.0* (pp. 109-121). IGI Global.

Mondal, S., Singh, S., & Gupta, H. (2023). Green entrepreneurship and digitalization enabling the circular economy through sustainable waste management-An exploratory study of emerging economy. *Journal of Cleaner Production*, 422, 138433. DOI: 10.1016/j.jclepro.2023.138433

Moşteanu, N. R. (2023). Navigating the intersection of sustainability and intelligent automation: Opportunities and challenges for entrepreneurs. *Human Resources Management and Services, 5*(1).

Muchenje, C., Mtengwa, E., & Maregere, L. L. (2024). Unleashing Entrepreneurial Potential: The Transformative Impact of Artificial Intelligence. In *AI-Driven Marketing Research and Data Analytics* (pp. 378-405). IGI Global.

Muriungi, K., & Chumo, K. P. (2024). Building Future-Proof, Sustainable, and Innovative Startups in Africa Powered by Smart Cities. In *Green Computing for Sustainable Smart Cities* (pp. 172–205). CRC Press. DOI: 10.1201/9781003388814-11

Niemi, M. (2024). Guiding Startups Towards Sustainability: A Qualitative Study on How Startups in the Turku Region Perceive Sustainability and External Support.

Paul, J., Lim, W. M., O'Cass, A., Hao, A. W., & Bresciani, S. (2021). Scientific procedures and rationales for systematic literature reviews (SPAR-4-SLR). *International Journal of Consumer Studies*, 45(4), O1–O16. DOI: 10.1111/ijcs.12695

Perifanis, N.-A., & Kitsios, F. (2023). Investigating the influence of artificial intelligence on business value in the digital era of strategy: A literature review. *Information (Basel)*, 14(2), 85. DOI: 10.3390/info14020085

Pizzi, S., Corbo, L., & Caputo, A. (2021). Fintech and SMEs sustainable business models: Reflections and considerations for a circular economy. *Journal of Cleaner Production*, 281, 125217. DOI: 10.1016/j.jclepro.2020.125217

Radonjić, A., Duarte, H., & Pereira, N. (2022). Artificial intelligence and HRM: HR managers' perspective on decisiveness and challenges. *European Management Journal.*

Razalli, N. M., Nashir, I. M., Hashim, M. E. A. H., Tahir, M. S. M., Othman, F. W., & Kleebrung, A. (2024). A Systematic Literature Review: Exploring the Integration of Technology in Green Entrepreneurship Education. *Journal of Advanced Research in Applied Sciences and Engineering Technology*, 248-264.

Saini, H., & Tarkar, P. (2022). Artificial intelligence in human resource practices with challenges and future directions. In *Handbook of Research on Innovative Management Using AI in Industry 5.0* (pp. 222–230). IGI Global. DOI: 10.4018/978-1-7998-8497-2.ch015

Sheng, T. J., Islam, M. S., Misran, N., Baharuddin, M. H., Arshad, H., Islam, M. R., Chowdhury, M. E. H., Rmili, H., & Islam, M. T. (2020). An internet of things based smart waste management system using LoRa and tensorflow deep learning model. *IEEE Access : Practical Innovations, Open Solutions*, 8, 148793–148811. DOI: 10.1109/ACCESS.2020.3016255

Tambe, P., Cappelli, P., & Yakubovich, V. (2019). Artificial intelligence in human resources management: Challenges and a path forward. *California Management Review*, 61(4), 15–42. DOI: 10.1177/0008125619867910

Todd, P., & Benbasat, I. (1991). An experimental investigation of the impact of computer-based decision aids on decision making strategies. *Information Systems Research*, 2(2), 87–115. DOI: 10.1287/isre.2.2.87

Van Eck, N. J., & Waltman, L. (2023). VOSviewer manual. Manual for VOSviewer version. 1.6.8.

Veleva, V. (2021). The role of entrepreneurs in advancing sustainable lifestyles: Challenges, impacts, and future opportunities. *Journal of Cleaner Production*, 283, 124658. DOI: 10.1016/j.jclepro.2020.124658 PMID: 33078048

Verma, R., & Bandi, S. (2019). Artificial intelligence & human resource management in Indian IT sector. *Proceedings of 10th international conference on digital strategies for organizational success.* DOI: 10.2139/ssrn.3319897

Vrontis, D., Christofi, M., Pereira, V., Tarba, S., Makrides, A., & Trichina, E. (2023). Artificial intelligence, robotics, advanced technologies and human resource management: a systematic review. *Artificial Intelligence and International HRM*, 172-201.

Wang, X., Zhang, J., Yao, K., & Qin, J. (2021). ResumeVis Interactive Visualization of Resumes Based on Multi-Source Data. *International Journal of Web Services Research*, 18(2), 40–53. DOI: 10.4018/IJWSR.2021040103

Zhang, J., & Chen, Z. (2024). Exploring human resource management digital transformation in the digital age. *Journal of the Knowledge Economy*, 15(1), 1482–1498. DOI: 10.1007/s13132-023-01214-y

Chapter 12
Sustainable Urban Mobility and Smart Traffic Management:
Balancing Optimization With Environmental Sustainability Goals

Saurabh Chandra
https://orcid.org/0000-0003-4172-9968
Bennett University, India

ABSTRACT

Urbanization and traffic congestion are rapidly worsening as an issue, and traffic optimization is necessary to improve mobility, security, and environmental sustainability. While there are diverse opportunities available, this chapter also identifies some emerging trends and technologies in traffic optimization, integration with smart city initiatives, long-term sustainability goals, etc. Advanced technologies like artificial intelligence (AI), integration of internet of things (IoT), and connected vehicles are changing traffic management by harnessing real-time data analysis and predictive routing. By embedding traffic strategies into smart cities, data-driven decision-making and public engagement leads to urban mobility improvements. Tying traffic management to sustainability objectives is also critical to reduce greenhouse gases and facilitate active modes of transportation. Through sustainable transportation, cities can improve transportation efficiency, address environmental issues, and enhance overall quality of life.

DOI: 10.4018/979-8-3693-9410-6.ch012

INTRODUCTION

Traffic optimization is an organized procedure that focuses on enhancing the effectiveness of road traffic movement, decreasing traffic congestion and going easy on the journey time. It is in this process where different technologies and methodologies are applied to control the movement of vehicular traffic, particularly in urban areas, where traffic congestion is a common problem (Dhananjay & Kumar, 2022). While the capabilities of these new technologies hold great promise, challenges still stand in the way, such as funding limitations, public acceptance, and technological limitations. To overcome these challenges, this paper suggests recommendations for policymakers and practitioners which focus on innovative funding mechanisms, engagement with communities, workforce development, and holistic regulatory frameworks. According to the Texas Transportation Institute, congestion causes individuals to spend, on average, 17 to 55 hours travel due to delays every year, emphasizing the need to identify effective traffic management solutions.

At the core of the goal of traffic optimization is the idea of improving the entire transportation experience by regressing the amount of time a vehicle spends waiting at stop lights, going through congested regions, and adapting to unpredictable traffic patterns. Real-time data analytics, adaptive signal control systems, and intelligent transportation systems (ITS) are all integral in this approach. Such technologies draw on data from multiple sources, including vehicle communication systems, cameras, and sensors to monitor traffic conditions and dynamically adapt operations. With the recent changes in the field of information technology have revolutionized the different techniques used in Traffic Optimization. Conventional methods were based on predetermined timings of traffic signals, which couldn't respond to the present state of flow. But the new approach uses adaptive traffic signal control systems that changes signal timing with real-time traffic flows. Such adaptability not only increases the number of vehicles able to traverse an area at any given time, but it also increases safety by making possibilities such as sudden stops or last-minute lane changes less likely to cause accidents (Ghosh & Mukherjee, 2023).

Significance of Clients Considering Environmental Aspects

Figure 1. Importance Environmental aspects

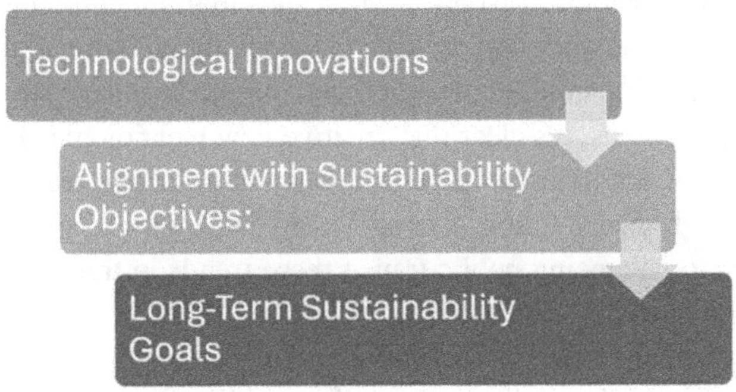

(source: author)

With the migration of people to urban areas and the growing number of vehicles each year, the environmental consequences of traffic management are becoming more critical. Congested traffic is not only a nuisance; it leads to more air pollution, more greenhouse gas emissions, and more fuel consumption. Therefore, it is important to incorporate ecological factors into optimization strategies for urban traffic. Effective traffic optimization has a host of environmental advantages, one of which is the reduction of greenhouse gas emissions. When Vehicles are running under congested situations, they release pollutants like carbon dioxide (CO_2). An increase in stop-and-go traffic can lead to higher pollution levels in a city, something that can be reduced if intelligent signals optimize traffic flow in real-time. Research indicates that such adaptive traffic management systems can reduce CO_2 by as much as 21% as they maximize vehicle movement efficiency3. Besides, making ways for better traffic also leads to better air quality in urban areas. High vehicular emissions are associated with respiratory diseases and other health problems in urban residents. Cities, therefore, may find themselves better positioned is in public health outcomes as well, in addition to meeting regulatory air quality standards, by reducing congestion and enabling smoother traffic patterns (Kuey & Chen, 2023).

Apart from cutting down emissions and boosting air quality, successful traffic optimization means less noise pollution, another major concern in urbanized environments. Road noise, it is pointed out, is often caused by sudden speed acceleration and deceleration associated with over-congested roads. Curbing noise pollution by adopting measures that encourage even driving speeds can improve the quality of life for residents living close to busy roads in a city.However, the economic ramifications of greener traffic optimization must be considered. But reducing congestion also saves time for commuters and reduces fuel consumption, which saves money for people and companies. Beyond individual savings, cities that prioritize efficient transportation systems tend to be more attractive to businesses and residents alike looking for a high quality of life (Pappas & Zervas, 2022). Traffic is always a problem, but the intersection of traffic optimization and environmental considerations is a critical area to try and work into policies to help with the more urban side of life. With the ever-increasing number of vehicles on our roads and the struggles against congestion in urban areas, it's clear that creative new technologies favouring both efficiency and eco-friendliness will be necessary in efforts to make urban spaces practical. From high-tech solutions like advanced data analytics to adaptive signal control systems, to enhancing public transit these trends in transportation will be the key to the future while combating inevitability of the environment. Types of such practices can help cities have cleaner communication networks, which will ultimately help around the globe combat climate change and move towards sustainable development. Traffic optimization is just one part of the equation, but in a time like this; with the pace of technological advancements, alongside a new wave of environmental consciousness, traffic optimization must remain at the forefront of urban planning programs (Ranjan & Singh, 2023).

Understanding Traffic Optimization

Traffic optimization refers to a set of approaches designed to enhance the performance of motor vehicles on road networks. Traffic optimization is fundamentally about minimizing delays, reducing congestion, and improving overall transportation efficiency. It's an area that involves the coordination of many different efforts focused on such things as the timing of signals, the use of travel lanes, route management and more. According to the Texas Transportation Institute, the average individuals suffer annual travel delays of 17 to 55 hours due to congestion, emphasizing the need for successful optimization approaches. Traffic optimization is not only a matter of speed but also of environmental harmony, safety, and economic efficiency (Shashikiran, Reddy, & Kumar, 2011). Cities which optimize traffic flow not only reduce greenhouse gas emissions but also improve air quality. In addition, better traffic control helps ensure an improved level of safety, as free flow lessens the chances of accidents that can occur in slower or jammed conditions. Traffic optimization refers to the ability to minimize traffic flow primarily through the most congested areas and improve overall route planning which applies to rural traffic optimization. With the advent of such advanced technologies like Intelligent Transportation Systems (ITS), the possibilities of optimizing the traffic in various contexts were increased. Integrated Traffic Systems (ITS) are designed to enhance real-time communication and information sharing between various transportation system components, enabling a more dynamic traffic control dynamic. (Raghavan & Reddy, 2023)

Key Technologies and Strategies

Critical Technologies and Approaches with the introduction of traffic optimization, a few technologies and strategies have become a significant part of it:

Adaptive Traffic Signal Control: Using real-time traffic data, this technology adjusts the timings of traffic signals to optimize traffic flow. Adaptive systems can optimize lengths of green light by data gleaned from sensors and cameras to break the momentum and maximize flow.

Traffic Flow Modeling Advanced modeling techniques simulate traffic patterns under various scenarios, allowing planners to anticipate the impact of changes in road infrastructure or traffic regulations. Such models are useful for identifying bottlenecks and estimating possible solutions (Bansal & Kaur, 2023) .

Connected vehicle technology: Cars and trucks with girds can communicate with one another and with infrastructure about their speed, location and direction. Traffic lighting can be optimized through this data, as well as real-time updates that provide drivers with information about congestion or alternative routes (Choudhury, 2023).

B. Real-Time Data Analytics: The real-time data that are generated from millions of vehicles enable continuous monitoring of traffic conditions. By studying historical and current data, cities can determine when to put money down on resources or adjust traffic management strategies (Davis, 2022).

Traffic Demand Management: These measures manage travel behavior, such as congestion pricing, which charges drivers for using specific roads during peak hours. Demand management reduces congestion by discouraging nonessential travel during busy times (Ibrahim, 2022).

Historical Context and Evolution

Integrating Public Transit: Efforts to optimize public transit routes and schedules can help minimize the number of vehicles on the road. (Finally, NUMO knows that with effective service to promote the use of public transportation, we can actually reduce general traffic volumes (Wang, Zhang, & Li, 2022).

Smart parking management: GCC cities have implemented technologies that provide real- time information on available parking spaces, and reduced the time taken by drivers looking for parking, which decreases the urban congestion (Hassan, 2023).

Context and Development Throughout History

The field of traffic optimization systems has evolved over the decades led by new technologies, new philosophies in urban planning, and environmental-awareness. Traffic management was historically focused on increasing roadway capacity through the development of infrastructure more lanes and new highways to carry more vehicles as they appeared. From the 1950s onward, cities started installing fixed-time signal systems based on fixed plans, tested against historical traffic patterns. While these systems yielded some improvements to flow, they were often rigid and incapable of adapting to real-time conditions. As cities grew and become more congested, it was clear that the one-size-fits-all approach was inadequate. The late 20th century brought another major breakthrough for traffic optimization: Computer technology. Advances in machine learning algorithms made it possible to analyze vast amounts of data, leading to more accurate prediction of traffic behaviors and the development of algorithms for assigning vehicle and pedestrian flows to intersections based on historical and real-time data. The availability of data has exploded in the 21st century, thanks to advances in sensor technology and mobile communications. Intelligent Traffic System (ITS) can be considered a new era in managing city traffic, as it provides active monitoring and reconfiguration of traffic systems. This move has enabled a more wholistic perspective on urban mobility that encompasses not just vehicle flow but also environmental impacts and public health outcomes (Fang, 2022).

Environmental Impacts of Traffic Optimization

Traffic optimization has been increasingly regarded as a necessity in combatting the environmental issues linked to urban transportation. Cities can greatly reduce greenhouse gas emissions, improve air quality, and help to decrease noise pollution simply by ensuring that traffic flows more efficiently. The rest of this section examines the three most critical environmental impacts of traffic optimization: greenhouse gas emissions, improvements in air quality, and reductions in noise pollution (Gonzalez, 2022).

Greenhouse Gas Emissions

Transportation is a leading driver of greenhouse gas (GHG) emissions, which accelerate climate change, contributing to extreme weather, drought, flooding, and ocean acidification affecting human health and the environment. About 33% of all carbon emissions in the nation come from transportation, most of which comes from fossil fuels burned in vehicles, according to the U.S. Department of Energy. Rising urban populations and increasing vehicle miles travelled (VMT) mean managing these emissions represents a growing challenge. Traffic optimization techniques can yield significant GHG reductions. As an example, adaptive traffic signal control systems could reduce up to 21% of vehicle emissions by reducing stop-and-go traffic patterns and improving overall traffic flow. These systems make use of real-time data to adapt the lengths of signal phases to the current traffic conditions, thus minimizing idle times and superfluous acceleration. In addition, a leading method is to utilize route optimization software, which enables drivers to choose the fastest and best routes. With minimized travel distances, and the avoidance of congested sectors, this technology can help reduce fuel consumption and cut CO2 output vastly. According to research, carbon emission can be reduced about 10% to 15%5 through route optimization In fact, research has found that smoothing out the flow of traffic using a host of interventions including congestion pricing and better management of vehicles on roads can reduce overall emissions from cars and trucks. For instance, the COSMO project, conducted in Europe, showed reductions in CO2 emissions between 5% and 15% by implementing innovative traffic management systems1. Not only do these reductions help mitigate climate change but they also create urban sustainability (Yang, Huang, & Li, 2023).

Air Quality Improvement

Traffic optimization has a deep connection with air quality. Vehicular emissions lead to poor air quality, which is a significant health risk linked with respiratory diseases and cardiovascular problems. In numerous metropolises, vehicular pollutants David giant main contributors of atmospheric commodities such as nitrogen oxides (NOx) and particulate matter (PM). Getting emissions down associated with congestion through traffic optimization can make a big difference in the quality of air. Vehicles release higher rates of pollutants when they are stuck in traffic or stop frequently and start again. Cities can reduce these harmful emissions by using strategies that improve flow on the surface streets, like adaptive signal control and real-time traffic management. For instance, London's congestion charge scheme has been correlated with better air quality results. Reduced congestion resulted in lower NOx and other pollutant levels and substantial public health benefits1 Likewise, studies undertaken in Stockholm suggested that interventions to reduce traffic congestion could avert many deaths from air pollution1. In addition, the integration of public transportation into traffic optimization can also improve air quality. In addition, cities can help the number of single-occupancy vehicles on United States roads by promoting the use of buses and trains that are duly scheduled and routed. Not only does this reduce overall emissions, it contributes to a healthier urban environment (Jiang, 2022).

Noise Pollution Reduction

Another major environmental impact of traffic is noise pollution. Human health is impacted as well as wildlife habitats. High road noise levels can turn into stress-subjective health disorders, sleep disturbances and decrease the life quality of the residents residing near roads. In addition, Optimization techniques for traffic can help reduce noise pollution, as it can help in reducing the heavy vehicles on the road and make traffic flow smoothly. Roads where vehicles can keep an even speed without having to decelerate often are less noisy than those with congested flow where rapid acceleration and brakes are the norm. Studies have shown that using tools like noise barriers, landscaping and low-noise pavements can help enhance noise mitigation and improvements2. Public transport vehicles can also benefit from fuel savings and lower marginal noise by route optimization. With GHG emissions and noise pollution, this is a game change when it comes to the adoption of electric vehicles (EVs) for urban fleets. EVs have zero tailpipe emissions and are much quieter than conventional internal combustion engine vehicles. Cities can play a role in substantial environmental benefits by promoting EV adoption through incentives and infrastructure development (such as charging stations) (Liu, 2023).

Traffic optimization has diverse but fundamentally interconnected environmental implications. So, with the right technologies and the right approach, traffic can be managed in a manner that more efficiently services the needs of the population, yet even reduces carbon emissions, as well as noise, and improves the quality of air in the urban environment. The advantages go beyond a greener planet but also result in improved public health. With the continuous growth of urban areas and the ever-increasing transportation challenges, focusing on traffic optimization will become of prime importance to reach sustainable development goals. Through innovative technologies that promote efficient mobility solutions, policymakers should understand how valuable environmental integration is in transportation planning processes. Different collaborators will need to put their heads together to maximise traffic optimization know-how Highlights traffic optimization know-how must be increasingly turned into actionable strategies to realise the full potential. Cities can create cleaner, safer and more liveable environments for own generations and the ones to come through collective efforts to improve urban mobility while minimizing the negative environmental impacts (Kumar, 2023).

Policies for Traffic Management

Urban mobility, safety and environmental footprint are only a few examples of areas in which traffic optimization can have an impact. From national policies to local government laws to international treaties, traffic optimization management needs an action framework that brings actions together in a comprehensive policy framework. Here we break down all three, their role, and how each connect to one another. (Amaral & Granath, 2016)

National Policies and Regulatons

Traffic optimization efforts across the country are primarily shaped and driven by the national policies. Federal transportation policy, which is created by federal agencies, provides guidelines, funds, and regulatory framework to support state and local action (Anthopoulos, 2017).

National Transportation Policy: So many federal national transportation policies focus on minimizing the infrastructure, public transport systems, and road safety in many countries. Examples of such systems can be found in the U.S. where the Federal Highway Administration (FHWA) drives the subject of Intelligent Transportation Systems (ITS). It is funded by the Federal Highway Administration (FHWA) for various projects within the Intelligent Transportation System (ITS) that potentially will lead to the improvement of traffic conditions on the roads by means of better management of public transport and reduction in travel times.

National Environmental Laws: National level laws such as environmental law often contain regulations which directly impact traffic management practices. A major example is the Clean Air Act in the U.S. that requires states to create State Implementation Plans (SIPs) to meet air quality standards. Our travel plans might also involve optimizing traffic through congestion pricing or improved public transit options to curb vehicle emissions (Axelsson & Granath, 2018).

Funding Programs Federal funding programs like the Transportation Investment Generating Economic Recovery (TIGER) grants back creative projects that can help improve traffic optimization. These programs incentivize local governments to implement technologies and strategies that reduce environmental impacts and enhance traffic flow.

Collaboration among Stakeholders: National policies also emphasize that data should be shared among different stakeholders in traffic management. These policies allow the creation of integrated systems through cooperation amongst federal agencies, state departments of transportation, and local governments that improve performance across districts (Chen, Zhang, & Wang, 2023).

Local Government Initiatives

They are also the ones at the frontiers of traffic optimization adjusted to their contexts, said the expert. They can also implement regulations, manage infrastructure and deploy new mobility technologies needed to address local traffic challenges.

Smart Traffic Signal Management: Other cities have developed advanced systems that adjust the traffic signals according to the real-time traffic conditions. Local authorities to purchase and install technologies that enable signal phases to adjust dynamically, thereby minimizing lags and enhancing overall traffic circulation (Zhao, Liu, & Wang, 2023).

Public Transport Improvements (Local projects often aim to accommodate public transport into traffic flow) Cities can establish dedicated bus lanes or prioritize transit vehicles at traffic signals to incentivize the use of public transport and discourage driving alone.

Congestion Pricing: There are some municipalities that have adopted congestion pricing models that charge drivers to use certain roadways during peak hours. This approach helps alleviate congestion by encouraging users to travel during off-peak periods or using other means of transportation.

Local governments are realizing the significance of community engagement in devising effective traffic optimization plans. Public consultations and participatory planning processes enable residents to raise concerns and propose improvements that can facilitate mobility within their neighbourhoods (Bhatia & Gupta, 2023).

Use Of Data: The local authorities rely on analytics to make a decision regarding traffic management. Or in the case of local governments, by each agency analysing historical and real-time data on traffic patterns, realizing and correcting bottlenecks and applying focused solutions to assist flow.

International Treaties and Norms

Global transportation challenges such as climate change and air pollution are increasingly addressed through international agreements and standards, which set a framework for cooperation between countries. Such agreements typically focus on sustainable development goals concerning urban mobility and environmental safeguard.

Sustainable Development Goals (SDGs): The UN's SDGs have targets around sustainable cities and communities (Goal 11). Effective urban transport planning and management practices help countries to build inclusive, safe, resilient and sustainable urban transport systems.

International Agreements on Climate: Policies coming from international agreements like the Paris Agreement illustrate the pressure on countries to reduce greenhouse gas emissions across all sectors, including transportation. Countries are encouraged to adopt policies that support sustainable transport solutions, such as traffic optimization that avoids the emission of pollutants.

ITF promotes some best practices among member countries regarding successful traffic optimization methods. Such forums provide a means by which countries can learn from one another, tailoring successful approaches to their respective contexts.

Technologies are Standardized; international standards bodies strive for ITS technologies applicable to traffic management. That's why the standardization is needed, it will help to ensure interoperability between different systems as well as to facilitate cross-border data sharing, which will promote the worldwide joint work on traffic flow optimization.Countries engage in collaborative research initiatives, which aim to develop innovative solutions to urban mobility challenges. Smart city tech, self-driving series and clean motoring are just a few examples of the focus of these partnerships–al of which can impact effective traffic optimization.

Traffic policy framework includes national, local and even global level policies for optimizing traffic. National policy got that all-important theoretical foundation, so that is really important, but then it also lays out this broad set of guideposts and funding opportunities for states and municipalities to basically come up with innovative solutions to their specific situations. Local governments are instrumental in implementing these plans through focused initiatives that improve traffic flow while respecting community needs. With urban population growth and the pressure on providing transportation-responsive solutions, a more comprehensive integration between domains would be required to holistically optimize the traffic system with minimal environmental concerns. Finding an optimal solution for the traffic problem and ensuring a smooth flow of goods requires continuous cooperation between various stakeholders, including government agencies at the national, regional and local levels, private sector partners to support digital and sustainable mobility solutions for all users, as well as community organizations and residents who can participate in the development of initiatives to provide more pleasant cities (Zolghadri-Jahromi & Zolghadri-Jahromi, 2022).

Technologies Standardization: ITS technologies standardization efforts rendered by the international standards organizations in traffic management If properly standardized this will allow interoperability of different systems and data-sharing across borders which boils down to more cooperation in optimizing traffic flows.

Cross-Border Research Collaborations: Countries often engage in cross-border research collaborations focused on creating novel solutions to urban mobility challenges. Such partnerships typically encompass smart city technologies, self-driving cars or sustainable transport any of which can lead to optimal traffic management.

Policy Framework for Traffic Optimization A multi-dimensional policy framework which involves national policies, local government initiatives and international agreements. National policies offer overarching guidelines and sources of funding that give states and municipalities the ability to adopt creative approaches to their distinct challenges. State and local governments implement these strategies by using emerging innovative micro-initiatives to optimize traffic flow while actively involving the community. Second, global pacts create a platform for dialogue and knowledge-sharing between nations on best practices for sustainable urban mobility. An interconnected policy framework will help implementation of the effective optimization framework in an efficient manner to reduce the environmental challenges. In the end, successful traffic management is about the continuous partnership between many different stakeholders—federal, state, and local government agencies; private sector partners; community organizations; and residents themselves and about their shared goals of traffic efficiency, environmental preservation, and quality of life improvement in urban areas (Ahmed & Khan, 2023).

Leveraging Data Analytics to Improve the Flow of Traffic

It provides insight into patterns and trends that aid in the decision-making process and is the key to an effective data analytics system that would optimize traffic flow. Cities can analyze massive amounts of data collected from various sources as a result of the integration of big data analytics into traffic management systems.

Historical Data Assessment: By analyzing historical traffic alongside real-time data cities can identify peak travel times, congestion patterns, and potential bottlenecks. Such an analysis allows transportation agencies to develop targeted interventions that mitigate congestion during peak demands.

Predictive Analytics: Using statistical algorithms and machine learning techniques, predictive analytics forecasts future traffic conditions based on historical data trends. This ability enables cities to anticipate congestion before it happens and allocate resources accordingly, like changing signal timings or rerouting vehicles.

Data-Driven Insights for Incident Management: Data analytics enhances incident management by providing real-time insights into accidents or breakdowns on the roadways. If incidents are reported quickly, emergency services can respond faster and traffic can be rerouted away from the effected areas, reducing interruptions.

Public Transport Optimization: When implemented in public transport systems, data analytics can be used to fine-tune routes and schedules based on demand trends. Transit agencies can then optimize the services by better meeting the passenger needs while further limiting delays by analyzing the ridership data.

Examples of Successful Implementations

Numerous cities worldwide have effectively implemented IT solutions using ITS and data analytics for traffic management. These case studies cover some of the best practices, as well as the outcomes realized using these technologies.

Los Angeles, CA: The City of Los Angeles has deployed a state-of-the-art adaptive signal control technology (ASCT) solution called the LA Traffic Management System (LATMS). Deploying real-time data from more than 4,000 traffic signals, this system optimizes signal timing based on current conditions. That has led to what LA has said is a big drop in the time it takes to get around up to 20% in peak hours and a reduction in car emissions, thanks to less idling.

Singapore: Singapore is well known for its all-encompassing smart city initiatives, its Intelligent Transport System (ITS) is widely acclaimed globally. Singapore controls congestion using a mixture of electronic road pricing (ERP) in conjunction with real-time traffic monitoring. The ERP system charges vehicles that enter congested regions during peak times, thereby encouraging off-peak travel and favouring public transportation. Consequently, Singapore has effectively kept to relatively low levels of traffic case, while improving its air quality.

Barcelona, Spain: The city of Barcelona has implemented a smart traffic management system that utilizes a variety of technologies, including IoT sensors and mobile applications. The system that the city operates displays real-time data regarding vacant parking, public transport schedules, and traffic conditions. With a reported 30% reduction in search time for parking spaces, this initiative has yielded better mobility throughout the city, resulting in lower congestion and emissions.

Glasgow, Scotland: Glasgow has an Intelligent Transport System that gives real-time public transport information to commuters through signage and mobile apps. Buses are outfitted with sensors that monitor their locations, so they can adjust to what passengers demand, dynamically. This has resulted in more reliable public transport and fewer private cars in the past (Alavi-Niaei & Hossainzadeh, 2023).

CONCLUSION

With the rapid course of urbanization and increased demands on transport systems, effective traffic optimization has become essential. In this in-depth investigation into traffic optimization, we have just seen how emerging technologies, smart city concepts, and long-term sustainability objectives have a great deal of common ground. It has further described the constraints and hurdles faced in the actualization of these strategies, lending a nuanced view of the context within which traffic optimization functions.

The implementation of Intelligent Transportation Systems (ITS), artificial intelligence, machine learning, and connected vehicle technologies are revolutionizing traffic management. Such innovations allow for real-time data analysis, improve predictive capabilities and enable dynamic routing that not only optimizes traffic flow, but also minimizes congestion. Successful traffic optimization is increasingly tied to larger smart city initiatives. Utilizing data-driven decision-making, engaging with the public, and using advanced simulation technologies such as digital twins can aid cities in producing more practical transportation solutions that are in-line with the needs of the community and help maximize mobility for all. Traffic optimization strategies should support long-term sustainability goals focused on minimizing greenhouse emissions, fostering active transportation options, and improving urban resiliency. To balance the consequences of urban mobility, cities should focus on environmentally responsible practices and incorporate green advancement choices into the traffic movement networks.Split urban areas often face several working challenges related to traffic optimization. Budget constraints limit the capacity of municipalities to invest in advanced technologies; social acceptance continues to be an impediment to the modification of behaviour; and technological constraints can inhibit the effective deployment of novel systems.

Recommendations for Policymakers and Practitioners

In order to harness the opportunities created at the juncture of traffic optimization and address existing challenges, some policy and practice recommendations are as follows:

Implement Funding Models: Policymakers should consider innovative funding models for traffic optimization initiatives. These may be in the form of public-private partnerships (PPPs), congestion pricing schemes, or dedicated taxes for transportation enhancements that directly support infrastructure and offer steady revenue streams.

Tell me about something that you can do in the future, what are your decisions you can make to promote public engagement? Such efforts will require public consultations and informed communication on the benefits of traffic optimization, as well as ongoing public outreach during proposed changes.

By investing in workforce development, you can ensure that your team has the skills and resources necessary to effectively manage new traffic management technologies. Governments must invest in training programs to provide junior staff with skills in data analytics, ITS deployment and system maintenance.

Data-Driven Decision-Making: Cities must adopt data-driven decision-making processes, supported by strong data collection and analysis systems. Data analytics, ideally with real-time data coming from many sources, can help design interventions that aim to improve the flow of traffic while also addressing local challenges.

Incorporate Multi-Modal Solutions: Traffic optimization initiatives should prioritize the development of multi-modal transportation solutions that promote the integration of multiple modes—like public transit, biking, walking, and ride-sharing—into interconnected markets. Encouraging integrated connectivity across these modes can improve total mobility while decreasing dependence on single-occupancy vehicle trips.

Define Concrete Sustainability Goals: Policymakers need to define concrete sustainability goals in relation to traffic optimization efforts. They have to set targets to reduce emissions, increase public transit ridership and promote active transportation options — and what should have been implemented well before October 2023. Periodic reviews of progress towards these goals will foster continuous improvement.

Promote Research Universities: Supporting research at universities or other advanced institutions is an important avenue for traffic innovation. Policymakers should work with academic institutions and private sector partners for to pursue new solutions that improve efficiency and address environmental issues.

Regulatory Frameworks: The regulation framework will help the traffic optimization planning process. Regulations should both encourage adherence to environmental standards and allow for innovations in deploying affordable technologies

Encourage Inter-Agency Collaboration at the Local Level: Traffic optimization efforts are most effective when they involve collaboration between local government agencies (such as the public safety, transportation, and planning departments), the private sector (including transportation service providers), community organizations, and residents. A community partnership allows a wider range of needs to be addressed and allows community resources to be maximized.

Tracking Results and Adjusting Tactics: Tracking results from traffic optimization is crucial for measuring effectiveness and allowing for real-time adjustments. Policymakers should create key performance indicators (KPIs) to assess progress toward congestion reduction, emissions mitigation, and improved mobility goals.

REFERENCES

Ahmed, F., & Khan, M. (2023). Smart public transport systems using IoT for efficient urban mobility. *Transportation Research Part D, Transport and Environment*, 116(1), 103682.

Alavi-Niaei, H., & Hossainzadeh. (2023). Machine learning applications in intelligent transportation systems: Current trends and future directions. *Artificial Intelligence Review*, 56(4), 103682.

Amaral, A., & Granath, J. (2016). Smart city initiatives: A framework for understanding urban innovation. *Journal of Urban Technology*, 23(3), 1–15. DOI: 10.1080/10630732.2016.1153134

Anthopoulos, L. (2017). Smart cities: A spatial approach to urban planning and development. *Sustainable Cities and Society*, 31, 1–12. DOI: 10.1016/j.scs.2017.02.001

Axelsson, K., & Granath, J. (2018). Reference architecture for context-aware intelligent traffic management systems in smart cities. *International Journal of Information Systems for Crisis Response and Management*, 10(4), 1–16. DOI: 10.4018/IJISCRAM.2018100101

Bansal, P., & Kaur. (2023). Traffic congestion prediction using deep learning techniques in smart cities. *Computers, Environment and Urban Systems*, 95(1), 101123.

Bhatia, S., & Gupta, S. (2023). AI-driven traffic management systems in smart cities: A review of current trends and future directions. *Sustainability*, 15(7), 5949. DOI: 10.3390/su15075949

Chen, Y., Zhang, X., & Wang, L. (2023). Optimizing traffic flow in smart cities: Soft GRU-based recurrent neural networks for enhanced congestion prediction using deep learning. *Sustainability*, 15(22), 15736. DOI: 10.3390/su152215736

Choudhury, R. D. (2023). Real-time adaptive traffic control systems based on AI techniques for urban environments. *Journal of Traffic Engineering*, 11(2), 102345.

Davis, L. W. (2022). The role of public engagement in developing smart city transportation solutions. *Urban Studies Journal*, 59(11), 2345–2360.

Dhananjay, M., & Kumar, S. (2022). Design and implementation of an ML and IoT based adaptive traffic management system for smart cities. *Journal of Traffic and Transportation Engineering*, 10(4), 301–315. DOI: 10.1016/j.jtte.2022.06.002

Fang, X. (2022). Enhancing urban mobility through integrated public transport systems: Lessons from global best practices. *Transport Policy Journal*, 97(1), 45–56.

Ghosh, S., & Mukherjee, A. (2023). Advances in the optimization of vehicular traffic in smart cities: Economic, social, and environmental impacts of AI technologies. *Sustainability*, 15(22), 15736. DOI: 10.3390/su152215736

Gonzalez, S. (2022). The impact of autonomous vehicles on urban traffic dynamics: Opportunities and challenges. *Transportation Research Part C, Emerging Technologies*, 128(1), 103–118.

Hassan, A. (2023). Leveraging big data analytics for intelligent transportation systems: A systematic review. *Transport Reviews Journal*, 43(3), 345–367.

Ibrahim, M. F. (2022). The future of mobility in smart cities: Trends and challenges aad. *Cities Journal*, 130(1), 103–114.

Jiang, Y. (2022). Evaluating the effectiveness of eco-friendly routing algorithms on urban air quality. *Environmental Science & Policy Journal*, 129(1), 67–78.

Kuey, C.-H., & Chen, Y.-C. (2023). AI-driven traffic management systems in smart cities: A review of current trends and future directions for sustainable urban mobility solutions. *Educational Administration: Theory and Practice*, 30(5), 105–116. DOI: 10.53555/kuey.v30i5.2780

Kumar, R. D. (2023). The role of machine learning in enhancing the efficiency of public transport networks. *Transportation Research Part A, Policy and Practice*, 162(1), 112–124.

Liu, Y. (2023). Innovations in real-time traffic monitoring technologies for sustainable urban development. *Sustainable Cities Journal*, 35(2), 145–158.

Pappas, M., & Zervas, P. E. (2022). Energy optimization for smart cities using IoT technologies: An integrated approach to urban energy management. *Energy Reports*, 8(5), 100–115.

Raghavan, V., & Reddy, K. V. (2023). Smart traffic management for sustainable development in cities: Enhancing safety and efficiency through technology. *Sustainable Cities and Society*, 85, 104–117.

Ranjan, P., & Singh, S. (2023). Real-time traffic management in smart cities: Insights from simulation and empirical data analysis. *Journal of Transportation Engineering*, 149(7), 04023015.

Shashikiran, S., Reddy, K. V., & Kumar, A. (2011). Optimal path selection based on traffic flow mapping using Kruskal's algorithm. *International Journal of Computer Applications*, 34(8), 1–5.

Wang, T., Zhang, D., & Li, H. (2022). A decentralized reinforcement learning-management system for efficient vehicular regulation in smart transportation systems. *IEEE Transactions on Intelligent Transportation Systems*, 23(9), 12012–12021.

Yang, Y., Huang, W., & Li, X. (2023). IoT-based real-time traffic management system for urban environments: A comprehensive review. *Sensors (Basel)*, 23(2), 1234–1256.

Zhao, Y., Liu, Y., & Wang, J. (2023). Blockchain technology in intelligent transportation systems: Enhancing security and transparency in traffic data management. *IEEE Access : Practical Innovations, Open Solutions*, 11(1), 2345–2358.

Zolghadri-Jahromi, M., & Zolghadri-Jahromi. (2022). Smart city applications for energy-efficient transportation systems: Challenges and opportunities. *Renewable & Sustainable Energy Reviews*, 153(20), 109210.

Compilation of References

. Panchuk, M., Kryshtopa, S., Sładkowski, A., & Panchuk, A. (2020). Environmental aspects of the production and use of biofuels in transport. Lecture Notes in Networks and Systems, 115–168. .DOI: 10.1007/978-3-030-42323-0_3

. Yuan, M. (2024). 15 Frontiers of GIScience. *Remote Sensing Handbook, Volume II: Image Processing, Change Detection, GIS, and Spatial Data Analysis*, 368.

Abbass, M., Abbas, U., Jafri, R., Arif, S. M., & Akhai, S. (2025). AI and Machine Learning Applications in Sustainable Smart Cities. In *Sustainable Smart Cities and the Future of Urban Development* (pp. 1–32). IGI Global Scientific Publishing.

Adams, R. (2024). The impact of autonomous vehicles on urban planning. *Journal of Urban Technology*, 31(2), 45–67. DOI: 10.1080/10630732.2024.1234567

Adams, R. (2025). Emerging trends in urban mobility post-pandemic: The rise of AVs. *Urban Studies Research.*

Ahmed, F., & Khan, M. (2023). Smart public transport systems using IoT for efficient urban mobility. *Transportation Research Part D, Transport and Environment*, 116(1), 103682.

Aishwarya, R. I. (2025). 28 Smart Cities for the Future. *Cybersecurity and Data Science Innovations for Sustainable Development of HEICC: Healthcare, Education, Industry, Cities, and Communities*, 406.

Akhavian, R., & Behzadan, A. H. (2016). Smartphone-based construction workers' activity recognition and classification. *Automation in Construction*, 71, 198–209. DOI: 10.1016/j.autcon.2016.08.015

Akram, M. W., Akram, N., Shahzad, F., Rehman, K. U., & Andleeb, S. (2024). Blockchain technology in a crisis: Advantages, challenges, and lessons learned for enhancing food supply chains during the COVID-19 pandemic. *Journal of Cleaner Production*, 434, 140034. DOI: 10.1016/j.jclepro.2023.140034

Al Halbusi, H., Popa, S., Alshibani, S. M., & Soto-Acosta, P. (2024). Greening the future: Analyzing green entrepreneurial orientation, green knowledge management and digital transformation for sustainable innovation and circular economy. *European Journal of Innovation Management*. Advance online publication. DOI: 10.1108/EJIM-02-2024-0169

Alateeg, S., Alhammadi, A., Al-Ayed, S. I., & Helmi, M. A. (2024). Factors influencing on behavioral intention to adopt artificial intelligence for startup sustainability. *Kurdish Studies*, 12(1), 2924–2941.

Alavi-Niaei, H., & Hossainzadeh. (2023). Machine learning applications in intelligent transportation systems: Current trends and future directions. *Artificial Intelligence Review*, 56(4), 103682.

Aldegheishem, A. (2023). Machine learning applications in smart cities: A case study of Riyadh. *Journal of Urban Technology*, 30(2), 145–162. DOI: 10.1080/10630732.2023.1234567

Alexopoulos, C., Saxena, S., Matheus, R., & Rizun, N. (2025). Artificial Intelligence (AI)-powered public service delivery in smart cities: The case of the Gulf Cooperation Council (GCC) Countries. In *ICT, the business sector and smart cities* (pp. 85-98). Routledge.

Ali, Z., Jamil, Y., Anwar, H., & Sarfraz, R. A. (2024). Classification of e-waste using machine learning-assisted laser-induced breakdown spectroscopy. Waste Management & Research.

Ali, J., & Khan, M. F. (2023). A trust-based secure parking allocation for IOT-enabled Sustainable Smart Cities. *Sustainability (Basel)*, 15(8), 6916. DOI: 10.3390/su15086916

Allam, Z., & Jones, D. S. (2020). On the coronavirus (COVID-19) outbreak and the smart city network: Universal data sharing standards coupled with artificial intelligence (AI) to benefit urban health monitoring and management. *Health Care*, 8(1), 46. DOI: 10.3390/healthcare8010046 PMID: 32120822

Almalawi, A., Zafar, A., Unhelkar, B., Hassan, S., Alqurashi, F., Khan, A. I., Fahad, A., & Alam, M. M. (2024). Enhancing security in smart healthcare systems: Using intelligent edge computing with a novel Salp Swarm Optimization and radial basis neural network algorithm. *Heliyon*, 10(13), e33792. DOI: 10.1016/j.heliyon.2024.e33792 PMID: 39040324

Al-Rimawi, T., & Nadler, M. (2025). Leveraging Smart City Technologies for Enhanced Real Estate Development: An Integrative Review. *Smart Cities*, 8(1), 10. DOI: 10.3390/smartcities8010010

Alsaeed, N., Nadeem, F., & Albalwy, F. (2024). A scalable and lightweight group authentication framework for Internet of Medical Things using integrated blockchain and fog computing. *Future Generation Computer Systems*, 151, 162–181. DOI: 10.1016/j.future.2023.09.032

Amaral, A., & Granath, J. (2016). Smart city initiatives: A framework for understanding urban innovation. *Journal of Urban Technology*, 23(3), 1–15. DOI: 10.1080/10630732.2016.1153134

Ameer, M., Rahul, S. P., & Manne, S. (2020). Human resource analytics using power bi visualization tool. 2020 4th international conference on intelligent computing and control systems (ICICCS).

Anand, A., Argade, P., Barkemeyer, R., & Salignac, F. (2021). Trends and patterns in sustainable entrepreneurship research: A bibliometric review and research agenda. *Journal of Business Venturing*, 36(3), 106092. DOI: 10.1016/j.jbusvent.2021.106092

Angelidou, M., Karachaliou, E., Angelidou, T., & Stylianidis, E. (2021). Cultural heritage in smart city environments. *ISPRS International Journal of Geo-Information*, 10(2), 71. DOI: 10.3390/ijgi10020071

Anthopoulos, L. (2017). Smart cities: A spatial approach to urban planning and development. *Sustainable Cities and Society*, 31, 1–12. DOI: 10.1016/j.scs.2017.02.001

Anthopoulos, L. G., & Reddick, C. G. (2020). Understanding electronic government research and smart city: A framework and empirical evidence. *Information Polity*, 25(2), 143–158. DOI: 10.3233/IP-190141

Appio, F. P., Platania, F., & Hernandez, C. T. (2024). Pairing AI and Sustainability: Envisioning Entrepreneurial Initiatives for Virtuous Twin Paths. *IEEE Transactions on Engineering Management*, 71, 11669–11686. DOI: 10.1109/TEM.2024.3428913

Arantes, L., & Costa, S. (2025). Converging Theories: The Fusion of Artificial Intelligence and Sustainable Marketing for a Responsible Future. In *Leveraging AI for Effective Digital Relationship Marketing* (pp. 223-256). IGI Global.

Arin, I. A., Dhamacora, R., & Aditya, B. (2024, September). The Adoption of Digital Twin Technology to Facilitate Resource and Asset Tracking in Field Service Management. In *2024 International Conference on ICT for Smart Society (ICISS)* (pp. 1-6). IEEE. DOI: 10.1109/ICISS62896.2024.10751066

Arora, M. K., Lal, S., Raghav, A., & Mishra, S. (2024). The Legal Implications of Corporate Borrowing and Analysing the Framework for Debt Financing and its Impact on Real Estate Sector in India. *National Journal of Real Estate Law*, 7(2), 1–4.

Asaithambi, S., Ravi, L., Devarajan, M., Almazyad, A. S., Xiong, G., & Mohamed, A. W. (2024). Enhancing enterprises trust mechanism through integrating blockchain technology into e-commerce platform for SMEs. *Egyptian Informatics Journal*, 25, 100444. DOI: 10.1016/j.eij.2024.100444

Ashraf, W. M., Jadhao, P. R., Panda, R., Pant, K. K., & Dua, V. (2024). Towards circular economy of wasted printed circuit boards of mobile phones fuelled by machine learning and robust mathematical optimization framework. *Resources, Conservation & Recycling Advances*, 23, 200226. DOI: 10.1016/j.rcradv.2024.200226

Axelsson, K., & Granath, J. (2018). Reference architecture for context-aware intelligent traffic management systems in smart cities. *International Journal of Information Systems for Crisis Response and Management*, 10(4), 1–16. DOI: 10.4018/IJISCRAM.2018100101

Aydinoglu, A. C., Bovkir, R., Colkesen, İ., Seyrek, E. C., & Bilgin, M. S. (2024). Developing a mobile application with geo-analytic tools to perform real estate valuation within smart cities. *Survey Review*, 1–19.

Baas, J., Schotten, M., Plume, A., Côté, G., & Karimi, R. (2020). Scopus as a curated, high-quality bibliometric data source for academic research in quantitative science studies. *Quantitative Science Studies*, 1(1), 377–386. DOI: 10.1162/qss_a_00019

Bable Smart Cities. (n.d.). AI for smart cities. Retrieved February 2, 2025, from https://www.bable-smartcities.eu/explore/solutions/solution/ai-for-smart-cities.html

Bag, S., Pretorius, J. H. C., Gupta, S., & Dwivedi, Y. K. (2021). Role of institutional pressures and resources in the adoption of big data analytics powered artificial intelligence, sustainable manufacturing practices and circular economy capabilities. *Technological Forecasting and Social Change*, 163, 120420. DOI: 10.1016/j.techfore.2020.120420

Baker, T. (2025). Innovations in public transportation systems with AV integration. *Journal of Public Transportation*.

Baker, T., & Smith, J. (2024). Cybersecurity challenges in autonomous driving systems. *International Journal of Automotive Technology*, 25(1), 15–29. DOI: 10.1007/s12239-024-00123-4

Bansal, P., & Kaur. (2023). Traffic congestion prediction using deep learning techniques in smart cities. *Computers, Environment and Urban Systems*, 95(1), 101123.

Belsare, K., Singh, M., Gandam, A., Malik, P. K., Agarwal, R., & Gehlot, A. (2024). An integrated approach of IoT and WSN using wavelet transform and machine learning for the solid waste image classification in smart cities. *Transactions on Emerging Telecommunications Technologies*, 35(4), e4857. DOI: 10.1002/ett.4857

Bhatia, S., & Gupta, S. (2023). AI-driven traffic management systems in smart cities: A review of current trends and future directions. *Sustainability*, 15(7), 5949. DOI: 10.3390/su15075949

Bibri, S. E. (2021). A novel model for data-driven smart sustainable cities of the future: From big data analytics to scientific holistic approaches. *Journal of Big Data*, 8(1), 1–53. DOI: 10.1186/s40537-021-00458-8 PMID: 33425651

Bocken, N. M. (2015). Sustainable venture capital–catalyst for sustainable start-up success? *Journal of Cleaner Production*, 108, 647–658. DOI: 10.1016/j.jclepro.2015.05.079

Bolívar, M. P. R. (2021). Governance models and outcomes to foster public value generation in smart cities. *Scandinavian Journal of Public Administration*, 25(1), 3–22.

Bolón-Canedo, V., Morán-Fernández, L., Cancela, B., & Alonso-Betanzos, A. (2024). A review of green artificial intelligence: Towards a more sustainable future. *Neurocomputing*, 599, 128096. DOI: 10.1016/j.neucom.2024.128096

Bonatti, P. A., Decker, S., Polleres, A., & Presutti, V. (2019). AIs: New directions for knowledge representation on the semantic web (dagstuhl seminar 18371). *Dagstuhl Reports*, 8(9).

Borkowski, P., & Kaczmarek, T. (2023). Dynamic multimodal transport systems with the participation of rail and sea transport. *Logistics (Basel)*, 8(2), 20. DOI: 10.2478/logi-2023-0020

Brenner, B. (2018). Transformative sustainable business models in the light of the digital imperative—A global business economics perspective. *Sustainability (Basel)*, 10(12), 4428. DOI: 10.3390/su10124428

Briz-Redón, Á., Martínez-Ruiz, F., & Montes, F. (2019). Spatial analysis of traffic accidents near and between road intersections in a directed Linear Network. Accident Analysis &amp. *Accident; Analysis and Prevention*, 132, 105252. DOI: 10.1016/j.aap.2019.07.028 PMID: 31437743

Broccardo, L., Zicari, A., Jabeen, F., & Bhatti, Z. A. (2023). How digitalization supports a sustainable business model: A literature review. *Technological Forecasting and Social Change*, 187, 122146. DOI: 10.1016/j.techfore.2022.122146

Brownfield, J., & White, M. (2023). The role of public-private partnerships in developing sustainable multimodal transport infrastructure. *International Journal of Project Management*, 41(2), 234–245.

Brown, T., & Greenfield, S. (2023). The role of policy frameworks in facilitating multimodal transportation integration: A comparative analysis of global practices. *Transport Reviews*, 43(2), 178–192.

Buchgeher, G., Gabauer, D., Martinez-Gil, J., & Ehrlinger, L. (2021). AIs in manufacturing and production: A systematic literature review. *IEEE Access : Practical Innovations, Open Solutions*, 9, 55537–55554. DOI: 10.1109/ACCESS.2021.3070395

Büchner, M., Zürn, J., Todoran, I. G., Valada, A., & Burgard, W. (2023). Learning and aggregating lane MLs for urban automated driving. In *Proceedings of the IEEE/CVF Conference on Computer Vision and Pattern Recognition* (pp. 13415-13424).

Budhwar, P., Malik, A., De Silva, M. T., & Thevisuthan, P. (2022). Artificial intelligence–challenges and opportunities for international HRM: A review and research agenda. *International Journal of Human Resource Management*, 33(6), 1065–1097. DOI: 10.1080/09585192.2022.2035161

Bumpus, A. (2019). Enabling digital energy entrepreneurship.

Burange, A. W., Deshmukh, V. M., Thakare, Y. A., & Shelke, N. A. (2025). Safeguarding the Internet of Things: Elevating IoT routing security through trust management excellence. *Computer Standards & Interfaces*, 91, 103873. DOI: 10.1016/j.csi.2024.103873

Cai, P., Wang, H., Sun, Y., & Liu, M. (2022). DQ-GAT: Towards safe and efficient autonomous driving with deep Q-learning and ML attention networks. *IEEE Transactions on Intelligent Transportation Systems*, 23(11), 21102–21112. DOI: 10.1109/TITS.2022.3184990

Caldarelli, G. (2024). Expert perspectives on blockchain in the circular economy: A Delphi study with industry specialists. *Journal of Cleaner Production*, 465, 142781. DOI: 10.1016/j.jclepro.2024.142781

Cao, J., Xu, Y., Yu, E., Xiang, Q., Song, K., He, L., & Cheng, G. (2024). GateKeeper: An UltraLite malicious traffic identification method with dual-aspect optimization strategies on IoT gateways. *Computer Networks*, 250, 110556. DOI: 10.1016/j.comnet.2024.110556

Carter, P., & Davis, N. (2023). Evaluating the effectiveness of mobility-as-a-service platforms in enhancing multimodal integration. *Journal of Transportation Engineering*, 149(6), 04023025.

Cavada, M., Hunt, D. V. L., & Rogers, C. D. F. (2020). Smart cities: Contradicting definitions and unclear measures. *Smart and Sustainable Built Environment*, 9(2), 73–91. DOI: 10.1108/SASBE-01-2019-0002

Chavda, B., Pillai, R., & Srivastava, A. (2025). Sustainable Migration: Unraveling the Connections Between Urban Planning, Migration, and AI. In *AI Strategies for Social Entrepreneurship and Sustainable Economic Development* (pp. 47-68). IGI Global Scientific Publishing.

Cheng, L., & Jin, J. (2020). Artificial intelligence and urban governance: A review of the current state of research and future prospects. *Cities (London, England)*, 102, 102763. DOI: 10.1016/j.cities.2020.102763

Cheng, W. (2024). Study on the Optimization of Urban Traffic and Spatial Structure in the Environment of Low-Carbon Eco-Cities. *International Journal of Knowledge Management*, 20(1), 1–14. DOI: 10.4018/IJKM.357086

Cheng, Z., Liu, Y., Wu, C., Pan, Y., Zhao, L., Deng, X., & Zhu, C. (2024). Decentralized IoT data sharing: A blockchain-based federated learning approach with joint optimizations for efficiency and privacy. *Future Generation Computer Systems*, 160, 547–563. DOI: 10.1016/j.future.2024.06.035

Chen, H., & Zhang, X. (2023). Evaluating the impact of integrated ticketing systems on user satisfaction in multimodal transport networks: Evidence from China's major cities. *Journal of Transport Geography*, 98, 103–115.

Chen, L. (2025). Assessing the environmental impacts of widespread AV adoption. *Environmental Science & Policy*.

Chen, L., & Wang, Y. (2024). Data privacy concerns in smart cities: A case study of autonomous vehicles. *Journal of Information Privacy and Security*, 20(3), 200–215. DOI: 10.1080/15536548.2024.1234568

Chen, W., Wu, W., Ouyang, Z., Fu, Y., Li, M., & Huang, G. Q. (2024). Event-based data authenticity analytics for IoT and blockchain-enabled ESG disclosure. *Computers & Industrial Engineering*, 190, 109992. DOI: 10.1016/j.cie.2024.109992

Chen, Y., Zhang, X., & Wang, L. (2023). Optimizing traffic flow in smart cities: Soft GRU-based recurrent neural networks for enhanced congestion prediction using deep learning. *Sustainability*, 15(22), 15736. DOI: 10.3390/su152215736

Chen, Z., Jiang, Y., & Sun, D. (2020). Discrimination and prediction of traffic congestion states of urban road network based on spatio-temporal correlation. *IEEE Access : Practical Innovations, Open Solutions*, 8, 3330–3342. DOI: 10.1109/ACCESS.2019.2959125

Chhtrapati, D., Chaudhari, S. P., Mevada, D., Bhatt, A., & Trivedi, D. (2021). Research productivity and network visualization on digital evidence: A bibliometric study. *Science & Technology Libraries*, 40(4), 358–372. DOI: 10.1080/0194262X.2021.1948486

Choudhury, R. D. (2023). Real-time adaptive traffic control systems based on AI techniques for urban environments. *Journal of Traffic Engineering*, 11(2), 102345.

Chowdhury, S., Dey, P., Joel-Edgar, S., Bhattacharya, S., Rodriguez-Espindola, O., Abadie, A., & Truong, L. (2023). Unlocking the value of artificial intelligence in human resource management through AI capability framework. *Human Resource Management Review*, 33(1), 100899. DOI: 10.1016/j.hrmr.2022.100899

Chu, T. T., & Pham, T. T. T. (2024). Vertical coordination in agri-food supply chain and blockchain: A proposed framework solution for Vietnamese cashew nut business. *Regional Science Policy & Practice*, 16(3), 12576. DOI: 10.1111/rsp3.12576

Cities, S. (2025). 7 AI-Optimized Urban Resource Management for Sustainable. *Financial Inclusion and Sustainable Development in Sub-Saharan Africa*, 96.

City of Nashville. (2023). Multimodal transportation analysis (MMTA) guideline. Nashville.gov. Retrieved from https://www.nashville.gov/departments/transportation/developers/multimodal-transportation-analysis

Cocchia, A., & Dameri, R. P. (2021). Exploring smart city vision and practice: A case study of Milan. *Journal of Urban Technology*, 28(1–2), 3–22. DOI: 10.1080/10630732.2020.1863905

Dai, S., Liang, Y., Liu, S., Wang, Y., Shao, W., Lin, X., & Feng, X. (2018, March). Learning entity and relation embeddings with entity description for AI completion. In *2018 2nd International Conference on Artificial Intelligence: Technologies and Applications (ICAITA 2018)* (pp. 194-197). Atlantis Press.

Dameri, R. P., & Rosenthal-Sabroux, C. (2020). Smart city and value creation: Towards a framework for public value co-creation. *International Journal of Knowledge-Based Development*, 11(1), 1–20. DOI: 10.1504/IJKBD.2020.105100

Dang, C., & Le, N. (2025). The Impact of Institutional Quality on the Relationships between Environmental Taxes, Carbon Dioxide Emissions, and Economic Growth in Developing Countries. DOI: 10.2139/ssrn.5110115

Das, D., Bapat, J., Katsenou, A., & Shrestha, S. (2025). Connectivity for AI enabled cities—A field survey based study of emerging economies. *arXiv preprint arXiv:2501.09479.*

Dasawat, S. S., & Sharma, S. (2023, May). Cyber security integration with smart new age sustainable startup business, risk management, automation and scaling system for entrepreneurs: an artificial intelligence approach. In *2023 7th international conference on intelligent computing and control systems (ICICCS)* (pp. 1357-1363). IEEE. DOI: 10.1109/ICICCS56967.2023.10142779

Davis, L. W. (2022). The role of public engagement in developing smart city transportation solutions. *Urban Studies Journal*, 59(11), 2345–2360.

Davis, M. (2025). Future directions for research on AV technology and urban design. *Journal of Urban Technology*.

Davis, M., & Thompson, R. (2024). The role of AI in enhancing traffic management systems. *Transportation Research Part C, Emerging Technologies*, 132, 102–115. DOI: 10.1016/j.trc.2023.102456

Davis, M., & Thompson, R. A. (2023). Enhancing user accessibility in multimodal transport systems through innovative design principles: Insights from Europe and North America. *Journal of Transport and Land Use*, 16(1), 123–140.

De Mauro, A., Greco, M., Grimaldi, M., & Ritala, P. (2018). Human resources for Big Data professions: A systematic classification of job roles and required skill sets. *Information Processing & Management*, 54(5), 807–817. DOI: 10.1016/j.ipm.2017.05.004

Deakin, M. (2020). Smart cities and the internet: From mode 2 to triple helix accounts of their evolution. *Journal of Urban Technology*, 27(1), 3–20. DOI: 10.1080/10630732.2019.1704205

Debnath, B., Chowdhury, R. R., Acharya, C., & Guha, A. (2024). Sustainability Aspects of E-waste Supply Chain Network Role of New and Emerging Technologies. Technological Advancement in E-waste Management, 63-79.

Dener, E., & Sutherland, C. (2021). Smart cities and the politics of urban data. *Urban Geography*, 42(3), 289–308. DOI: 10.1080/02723638.2020.1796046

Dey, D., Majumder, A., Agrawal, Y., Tewari, S., & Mohapatra, H. (2025). Smart Mobility Revolution: Harnessing IoT, Sensors, and Cloud Computing for Intelligent Automobiles in the Urban Landscape. In *Sustainable Smart Cities and the Future of Urban Development* (pp. 143-164). IGI Global Scientific Publishing.

Dhananjay, M., & Kumar, S. (2022). Design and implementation of an ML and IoT based adaptive traffic management system for smart cities. *Journal of Traffic and Transportation Engineering*, 10(4), 301–315. DOI: 10.1016/j.jtte.2022.06.002

Díaz-Díaz, R., Muñoz, L., & Pérez-González, D. (2020). Business model analysis of public services operating in the smart city ecosystem: The case of SmartSantander. *Future Generation Computer Systems*, 108, 1052–1063. DOI: 10.1016/j.future.2017.05.041

Diaz-Llairo, M. A. (2025). Exploring a Smart City and the Epicenter of Megatrends in Innovation With Artificial Intelligence: The Case of Dubai. In *Data-Driven Governance Through AI, Digital Marketing, and the Privacy Interplay* (pp. 191-238). IGI Global Scientific Publishing.

Dizon, E., & Pranggono, B. (2021). Smart streetlights in smart city: A case study of Sheffield. *Journal of Ambient Intelligence and Humanized Computing*, 13(4), 2045–2060. DOI: 10.1007/s12652-021-02970-y

Dong, K., Ni, G., Taghizadeh-Hesary, F., & Zhao, C. (2023). Does smart transportation matter in inhibiting carbon inequality. *Energy Economics*, 126, 106952. DOI: 10.1016/j.eneco.2023.106952

Dong, X. L., He, X., Kan, A., Li, X., Liang, Y., Ma, J., & Han, J. (2020, August). Autoknow: Self-driving knowledge collection for products of thousands of types. In *Proceedings of the 26th ACM SIGKDD International Conference on Knowledge Discovery & Data Mining* (pp. 2724-2734). DOI: 10.1145/3394486.3403323

Donthu, N., Kumar, S., Mukherjee, D., Pandey, N., & Lim, W. M. (2021). How to conduct a bibliometric analysis: An overview and guidelines. *Journal of Business Research*, 133, 285–296. DOI: 10.1016/j.jbusres.2021.04.070

Dutta, P. K., Singh, B., Lal, S., Arora, M. K., Raghav, A., Mensah, G. B., & Alkattan, H. (2023). Encoding IoT for patient monitoring and smart healthcare: Connected healthcare system fostering health 6.0. *Babylonian Journal of Internet of Things*, 2023, 48–58. DOI: 10.58496/BJIoT/2023/007

Dutta, P. K., Singh, B., Towfeek, A.-S. K., Adamopoulou, J. P., Bardavouras, A. N., Bamwerinde, W., Turyasingura, B., & Ayiga, N. (2024). IoT Revolutionizes Humidity Measurement and Management in Smart Cities to Enhance Health and Wellness. *Mesopotamian Journal of Artificial Intelligence in Healthcare*, 2024, 110–117. DOI: 10.58496/MJAIH/2024/013

Edwards, K., & Johnson, P. (2024). Public perception of autonomous vehicles: A survey analysis. *Journal of Transport and Land Use*, 17(1), 89–105. DOI: 10.5198/jtlu.v17i1.1234

Ejaz, U., Ramon, W., & Olaoye, G. (2025). The Role of Big Data and AI in Smart Cities and Urban Planning.

Elahi, M. F., Luo, X., & Tian, R. (2020, July). A framework for modeling AIs via processing natural descriptions of vehicle-pedestrian interactions. In *International Conference on Human-Computer Interaction* (pp. 40-50). Cham: Springer International Publishing.

Elsayed, A., & Mohamed, M. (2025). Enhancing Smart City Management with AI: Analyzing Key Criteria and their Interrelationships using DEMATEL under Neutrosophic Numbers and MABAC for Optimal Development. *Neutrosophic Systems with Applications*, 25(2), 1–38. DOI: 10.61356/j.nswa.2025.25480

Eremia, M., Toma, L., & Sanduleac, M. (2020). The smart city concept in the 21st century. *Procedia Engineering*, 181, 12–19. DOI: 10.1016/j.proeng.2017.02.357

Eriksson, R. O., & Regoczi, B. (2023). Diving into venture creation: An exploratory study of how external enabler combinations facilitate sustainable venture creation in the marine sector.

Etensa, T., Alemu, T., & Yayo, M. (2025). Rethinking the measurements and predictors of environmental degradation in Ethiopia: Predicting long-term impacts using a kernel-based machine learning approach. *Environmental and Sustainability Indicators*, 25, 100554. DOI: 10.1016/j.indic.2024.100554

Fang, X. (2022). Enhancing urban mobility through integrated public transport systems: Lessons from global best practices. *Transport Policy Journal*, 97(1), 45–56.

Farooq, M., Ramzan, M., & Yen, Y. Y. (2025). Exploring the Impact of Artificial Intelligence on Brand Management.

Farshadfar, Z., Mucha, T., & Tanskanen, K. (2024). Leveraging Machine Learning for Advancing Circular Supply Chains: A Systematic Literature Review. *Logistics (Basel)*, 8(4), 108. DOI: 10.3390/logistics8040108

Fattah, M., Morshed, S. R., & Kafy, A.-A. (2022). Insights into the socio-economic impacts of traffic congestion in the port and industrial areas of Chittagong City, Bangladesh. *Transportation Engineering*, 9, 100122. DOI: 10.1016/j.treng.2022.100122

Force, I. A. I. T. (2022). A review of the methods, applications, and challenges of adopting artifical intelligence in the property assessment office. *Journal of Property Tax Assessment & Administration*, 19(1), 2.

Franqa, A., & Ullah, M. (2020). Integrating machine learning for the sustainable development of smart cities. *Frontiers in Sustainable Cities*, 4, 1449404. Advance online publication. DOI: 10.3389/frsc.2024.1449404

Füller, J., Hutter, K., Wahl, J., Bilgram, V., & Tekic, Z. (2022). How AI revolutionizes innovation management–Perceptions and implementation preferences of AI-based innovators. *Technological Forecasting and Social Change*, 178, 121598. DOI: 10.1016/j.techfore.2022.121598

Fu, X., Wang, L., Yuan, L., Hu, H., Li, T., Zhang, J., Ke, Y., Wang, M., Gao, Y., Huo, W., Chen, Y., Zhang, W., Liu, J., Huang, Z., Zhao, Y., Hu, F., Zhang, M., Liu, Y., Sun, X., & Hu, D. (2023). Long-term exposure to traffic noise and risk of incident cardiovascular diseases: A systematic review and dose-response meta-analysis. *Journal of Urban Health*, 100(4), 788–801. DOI: 10.1007/s11524-023-00769-0 PMID: 37580544

Garcia, S., & Martinez, A. (2024). Regulatory frameworks for autonomous vehicle deployment. *Policy Studies Journal: the Journal of the Policy Studies Organization*, 52(2), 321–340. DOI: 10.1111/psj.12456

Ghazali, W. N., Zulkifli, C. N., & Ponrahono, Z. (2019). The effect of traffic congestion on quality of Community Life. The European Proceedings of Multidisciplinary Sciences, 759–766. .DOI: 10.15405/epms.2019.12.77

Giffinger, R., & Gudrun, H. (2020). Smart cities ranking: An effective instrument for the positioning of the cities? *ACE: Architecture. City and Environment*, 4(12), 7–25. DOI: 10.5821/ace.v4i12.2483

Gil-Garcia, J. R., Pardo, T. A., & Nam, T. (2020). What makes a city smart? Identifying core components and proposing an integrative and comprehensive conceptualization. *Information Polity*, 25(1), 1–19. DOI: 10.3233/IP-190126

Giuggioli, G., & Pellegrini, M. M. (2023). Artificial intelligence as an enabler for entrepreneurs: A systematic literature review and an agenda for future research. *International Journal of Entrepreneurial Behaviour & Research*, 29(4), 816–837. DOI: 10.1108/IJEBR-05-2021-0426

Giuggioli, G., Pellegrini, M. M., & Giannone, G. (2024). Artificial intelligence as an enabler for entrepreneurial finance: A practical guide to AI-driven video pitch evaluation for entrepreneurs and investors. *Management Decision*. Advance online publication. DOI: 10.1108/MD-10-2023-1926

Gnanasekaran, C., Govindaraj, M., Jayavelu, S., & Chandrasekar, V. (2024, November). IoT and strategic management in action: Pilot testing for efficient e-waste collection in rural Indian households. In AIP Conference Proceedings (Vol. 3192, No. 1). AIP Publishing.

Gokulapriyan, R., Vinod, L., Chandran, R. A., & Yedla, V. R. (2025). Smart City Projects Influencing Smart Tourist Destinations. In *Advancing Smart Tourism Through Analytics* (pp. 333–350). IGI Global Scientific Publishing.

Gong, T., Jiang, Y., Saldivia, L. E., & Agard, C. (2022). Using Sankey diagrams to visualize drag and drop action sequences in technology-enhanced items. *Behavior Research Methods*, 54(1), 117–132. DOI: 10.3758/s13428-021-01615-4 PMID: 34109559

Gonzalez, S. (2022). The impact of autonomous vehicles on urban traffic dynamics: Opportunities and challenges. *Transportation Research Part C, Emerging Technologies*, 128(1), 103–118.

Grabocka, E., & Ndoka, E. (2025). AI-driven innovation within the ICT sector. *Smart Cities and Regional Development (SCRD)Journal*, 9(1), 77–97.

Gregori, P., & Holzmann, P. (2020). Digital sustainable entrepreneurship: A business model perspective on embedding digital technologies for social and environmental value creation. *Journal of Cleaner Production*, 272, 122817. DOI: 10.1016/j.jclepro.2020.122817

Griffin, A. L., Reichenbacher, T., Liao, H., Wang, W., & Cao, Y. (2024). Cognitive issues of mobile map design and use. *Journal of Location Based Services*, 18(4), 1–31. DOI: 10.1080/17489725.2024.2371288

Grubmüller, B. (2023). Analysis of startup stories through the lens of the external enabler framework.

Güngör, Z., Serhadlıoğlu, G., & Kesen, S. E. (2009). A fuzzy AHP approach to personnel selection problem. *Applied Soft Computing*, 9(2), 641–646. DOI: 10.1016/j.asoc.2008.09.003

Guo, R., Hong, Z., & Xue, X. (2021). *Reward Function Design via Human AI and Inverse Reinforcement Learning for Intelligent Driving* (No. 2021-01-0180). SAE Technical Paper.

Gupta, M., & Bhatnagar, R. (2025). Harnessing AI in Geospatial Technology for Environmental Monitoring and Management: Applications of AI for Geospatial Data Processing. In *Recent Trends in Geospatial AI* (pp. 73-102). IGI Global Scientific Publishing.

Gupta, M., Deswal, P., Chauhan, V., Kumar, P., Kaur, J., & Singh, P. (2025). Optimizing Sustainability: Harnessing AI for Personalized Electric Vehicle Energy Management. In *Optimized Energy Management Strategies for Electric Vehicles* (pp. 233-260). IGI Global Scientific Publishing.

Gupta, S., Dwivedi, A., & Kumar, H. (2025). Digital Platforms as Catalysts of Circular Economy in Pursuit of SDG 11: A Rigorous Exegesis of Smart Cities. In *Innovating Sustainability Through Digital Circular Economy* (pp. 373-390). IGI Global Scientific Publishing.

Hadizadeh, M., Ghaffari Feyzabadi, J., Fardi, Z., Mortazavi, S. M., Braga, V., & Salamzadeh, A. (2024). Digital Platforms as a Fertile Ground for the Economic Sustainability of Startups: Assaying Scenarios, Actions, Plans, and Players. *Sustainability (Basel)*, 16(16), 7139. DOI: 10.3390/su16167139

Hajoary, P. K., Jennifer, D. S., & Pathak, M. (2024). Digital technology adoption in circular startups: An integrated framework. *Business Strategy & Development*, 7(3), e425. DOI: 10.1002/bsd2.425

Halilaj, L., Dindorkar, I., Lüttin, J., & Rothermel, S. (2021). A AI-based approach for situation comprehension in driving scenarios. *The Semantic Web: 18th International Conference, ESWC 2021, Virtual Event, June 6–10, 2021Proceedings*, 18, 699–716.

Halilaj, L., Luettin, J., Henson, C., & Monka, S. (2022, September). AIs for automated driving. In *2022 IEEE Fifth International Conference on Artificial Intelligence and Knowledge Engineering (AIKE)* (pp. 98-105). IEEE.

Halilaj, L., Luettin, J., Monka, S., Henson, C., & Schmid, S. (2023). AI-Based Integration of Autonomous Driving Datasets. *International Journal of Semantic Computing*, 17(02), 249–271. DOI: 10.1142/S1793351X23600048

Hall, E., & Roberts, L. (2024). Sustainable urban mobility: Integrating electric and autonomous vehicles. *Urban Studies (Edinburgh, Scotland)*, 61(5), 1023–1040. DOI: 10.1177/0042098023123456

Hamadeh, N., Karouni, A., & Farhat, Z. (2022). Intelligent Transportation Systems to mitigate road traffic congestion. *Intelligenza Artificiale*, 15(2), 91–104. DOI: 10.3233/IA-200079

Hammouch, H., & Raghav, A. (2025). AI and ML Applications in Wildlife Conservation and Forest Management: A Comprehensive Review. *Machine Learning and Internet of Things in Fire Ecology*, 49-68.

Hansen, M. M., Koonsanit, K., & Kulmala, V. (2025). How can data contribute to Smart City innovation: A study from Thailand's Smart City initiatives. *Frontiers in Sustainable Cities*, 6, 1473123. DOI: 10.3389/frsc.2024.1473123

Hasan, H. R., Musamih, A., Salah, K., Jayaraman, R., Omar, M., Arshad, J., & Boscovic, D. (2024). Smart agriculture assurance: IoT and blockchain for trusted sustainable produce. *Computers and Electronics in Agriculture*, 224, 109184. DOI: 10.1016/j.compag.2024.109184

Hasan, M., Rahman, M. S., Janicke, H., & Sarker, I. H. (2024). Detecting anomalies in blockchain transactions using machine learning classifiers and explainability analysis. *Blockchain: Research and Applications*, 100207(3), 100207. Advance online publication. DOI: 10.1016/j.bcra.2024.100207

Hassan, A. (2023). Leveraging big data analytics for intelligent transportation systems: A systematic review. *Transport Reviews Journal*, 43(3), 345–367.

Hernandez, F. J., & Lopez, C. (2023). Integrating environmental sustainability into multimodal transportation planning: A framework for decision-making. *Sustainable Cities and Society*, 85(1), 104032.

Hind, H., Lal, S., & Arora, M. K. (2025). Artificial Intelligence and Urban Sustainability. *Developing Smart Sustainability With Artificial Intelligence*, 17.

Hmoud, B., & Varallyai, L. (2023). Role of Artificial Intelligence in Human Resource Management in the Middle East Countries. *KnE Social Sciences*, 435–448.

Hollands, R. G. (2020). Will the real smart city please stand up? Intelligent, progressive or entrepreneurial? *City (London, England)*, 12(3), 303–320. DOI: 10.1080/13604810802479126

Holzinger, A., Saranti, A., Hauschild, A. C., Beinecke, J., Heider, D., Roettger, R., . . . Pfeifer, B. (2023, August). Human-in-the-loop integration with domain-AIs for explainable federated deep learning. In *International Cross-Domain Conference for Machine Learning and Knowledge Extraction* (pp. 45-64). Cham: Springer Nature Switzerland. DOI: 10.1007/978-3-031-40837-3_4

Huang, H., Sun, L., Du, B., Liu, C., Lv, W., & Xiong, H. (2021, August). Representation learning on AIs for node importance estimation. In *Proceedings of the 27th ACM SIGKDD Conference on Knowledge Discovery & Data Mining* (pp. 646-655).

Hung, T. H., Pandey, A., Mittal, S., & Chandel, A. (2025). Harnessing IoT and geomarketing for agile marketing. In *Digital Transformation Initiatives for Agile Marketing* (pp. 167–186). IGI Global.

Hurbean, L., & Mishra, A. (2021). Future of smart cities: The role of machine learning and artificial intelligence. *Smart City Journal*, 5(3), 110–119. DOI: 10.2139/smartcityjournal.v5i3.110

Hu, X., Liu, Y., Tang, B., Yan, J., & Chen, L. (2023). Learning dynamic ML for overtaking strategy in autonomous driving. *IEEE Transactions on Intelligent Transportation Systems*, 24(11), 11921–11933. DOI: 10.1109/TITS.2023.3287223

Ibrahim, M. F. (2022). The future of mobility in smart cities: Trends and challenges aad. *Cities Journal*, 130(1), 103–114.

Institute, G. S. V. (2024). MBA in multi-modal transportation. Retrieved from https://gsv.ac.in/course/mba-multi-modal-transportation/

Ivković, R., Petrović, M., Spalević, P., & Milivojević, Z. (2024). Classification of E-Waste Types Using Machine Learning and Digital Image Processing. *Przeglad Elektrotechniczny*, 2024(9).

Iyer, R., & Patel, S. (2024). Ethical considerations in the deployment of autonomous vehicles. *Ethics and Information Technology*, 26(1), 55–67. DOI: 10.1007/s10676-023-09612-w

Jadaan, K., Alsarayreh, D., & Obaid, M. (2021). Observing people's reactions and responses to urban road traffic noise (RTN) in Jordan. Journal of Civil Engineering. *Science and Technology*, 12(2), 203–212. DOI: 10.33736/jcest.3984.2021

Jain, R. (2025). Integrating Artificial Intelligence for Smart Destination Management in the UAE: Opportunities, Challenges, and Impacts. *Advancing Smart Tourism Through Analytics*, 245-260.

Javaid, M., Haleem, A., Singh, R. P., Suman, R., & Gonzalez, E. S. (2022). Understanding the adoption of Industry 4.0 technologies in improving environmental sustainability. *Sustainable Operations and Computers*, 3, 203–217. DOI: 10.1016/j.susoc.2022.01.008

Javeed, D., Saeed, M. S., Ahmad, I., Adil, M., Kumar, P., & Islam, A. K. M. N. (2024). Quantum-empowered federated learning and 6G wireless networks for IoT security: Concept, challenges and future directions. *Future Generation Computer Systems*, 160, 577–597. DOI: 10.1016/j.future.2024.06.023

Jayawardena NS, Behl A, Thaichon P, & Quach S (2022). Artificial intelligence (AI)-based market intelligence and customer insights. *Artificial Intelligence for Marketing Management*, 120-41.

Jha, A., Sindhwani, R., Dwivedi, A., & Saddikuti, V. (2022). Sustainable recovery for digital entrepreneurs with shared resources: Enablers, challenges and solutions. *Journal of Asia Business Studies*, 16(3), 515–537. DOI: 10.1108/JABS-05-2021-0214

Jiang, Y. (2022). Evaluating the effectiveness of eco-friendly routing algorithms on urban air quality. *Environmental Science & Policy Journal*, 129(1), 67–78.

Ji, S., Pan, S., Cambria, E., Marttinen, P., & Philip, S. Y. (2021). A survey on AIs: Representation, acquisition, and applications. *IEEE Transactions on Neural Networks and Learning Systems*, 33(2), 494–514. DOI: 10.1109/TNNLS.2021.3070843 PMID: 33900922

Johnson, M., & Lee, H. (2024). Smart infrastructure for autonomous vehicles: Challenges and opportunities. *Journal of Infrastructure Systems*, 30(2), 45–60. DOI: 10.1061/(ASCE)IS.1943-555X.00001234

Jordan, M. I., & Mitchell, T. M. (2015). Machine learning: Trends, perspectives, and prospects. *Science*, 349(6245), 255–260. DOI: 10.1126/science.aaa8415 PMID: 26185243

Jorzik, P., Antonio, J. L., Kanbach, D. K., Kallmuenzer, A., & Kraus, S. (2024). Sowing the seeds for sustainability: A business model innovation perspective on artificial intelligence in green technology startups. *Technological Forecasting and Social Change*, 208, 123653. DOI: 10.1016/j.techfore.2024.123653

Jose, S. (2019). Innovation in recruitment and talent acquisition: A study on technologies and strategies adopted for talent management in IT sector. *International Journal of Marketing and Human Resource Management*, 10(2), 1–8. DOI: 10.34218/IJMHRM.10.3.2019.001

Joshi, P., Bagul, P., & Deshpande, A. (2024, October). Towards Sustainable E-Waste Management: A Computer Vision and Machine Learning Approach for Identification and Valuation. In *2024 International Conference on Intelligent Systems and Advanced Applications (ICISAA)* (pp. 1-6). IEEE. DOI: 10.1109/ICISAA62385.2024.10828717

Kalawatia, M., & Rajeraman, B. (2024). Technology and shared mobility: Accelerating multimodal integration in Indian cities through open networks. *Observer Research Foundation*. Retrieved from https://www.orfonline.org/research/shared-mobility-india

Kaldeli, E., Menis-Mastromichalakis, O., Bekiaris, S., Ralli, M., Tzouvaras, V., & Stamou, G. (2021). CrowdHeritage: Crowdsourcing for improving the quality of cultural heritage metadata. *Information (Basel)*, 12(02), 64. DOI: 10.3390/info12020064

Kamandang, Z. R., Wibisana, H., & Casita, C. B. (2020). Analysis of noise pollution due to traffic volume based on the types of automotive vehicles: A study of west surabaya region. *Journal of Physics: Conference Series*, 1569(4), 042044. DOI: 10.1088/1742-6596/1569/4/042044

Kaur, R., & Singh, P. (2022). Smart transportation systems powered by machine learning: A review of applications and challenges. *International Journal of Transportation Science and Technology*, 11(1), 25–40. DOI: 10.1016/j.tstl.2022.01.002

Kaushal, N., Kaurav, R. P. S., Sivathanu, B., & Kaushik, N. (2023). Artificial intelligence and HRM: Identifying future research Agenda using systematic literature review and bibliometric analysis. *Management Review Quarterly*, 73(2), 455–493. DOI: 10.1007/s11301-021-00249-2

Kaushik, T. K., Singh, A., Raghav, A., & Singh, B. (2024). Potential of Artificial Intelligence in the Agricultural Industry: A Comprehensive Review. In B. Riswandi, B. Singh, C. Kaunert, & K. Vig (Eds.), AI Applications for Clean Energy and Sustainability (pp. 344-365). IGI Global. DOI: 10.4018/979-8-3693-6567-0.ch017

Kaushik, T. K., Singh, A., Raghav, A., Singh, B., Arora, M. K., & Lal, S. (2024). Untapping Artificial Intelligence in Environmental and Sustainable Progression: Projecting Legal and Ethical Considerations. In B. Singh, C. Kaunert, K. Vig, & S. Dutta (Eds.), *Maintaining a Sustainable World in the Nexus of Environmental Science and AI* (pp. 109-130). IGI Global. DOI: 10.4018/979-8-3693-6336-2.ch005

Kaushik, T. K., Singh, A., Raghav, A., Singh, B., Arora, M. K., & Lal, S. (2024). Untapping Artificial Intelligence in Environmental and Sustainable Progression: Projecting Legal and Ethical Considerations. In *Maintaining a Sustainable World in the Nexus of Environmental Science and AI* (pp. 109-130). IGI Global.

Khan, S. (2020). Artificial Intelligence Virtual Assistants (Chatbots) are Innovative Investigators. *IJCSNS, 20*(2).

Khan, I. U., Taherdoost, H., Madanchian, M., Ouaissa, M., El Hajjami, S., & Rahman, H. (Eds.). (2024). *Future Tech Startups and Innovation in the Age of AI*. CRC Press.

Khazaei, P. (n.d.). Evaluating the Role of Artificial Intelligence in Crafting Emotional Advertising and Enhancing Customer Experience in Diverse Industry Sectors.

Kiani, A. (2024). Artificial intelligence in entrepreneurial project management: A review, framework and research agenda. *International Journal of Managing Projects in Business*. Advance online publication. DOI: 10.1108/IJMPB-03-2024-0068

Kim, J. E., Henson, C., Huang, K., Tran, T. A., & Lin, W. Y. (2021). Accelerating road sign ground truth construction with AI and machine learning. In *Intelligent Computing:Proceedings of the 2021 Computing Conference,* Volume 2 (pp. 325-340). Springer International Publishing.

Kim, J. H., & Lee, S. M. (2023). The impact of real-time travel information on user satisfaction in multimodal transport networks. *Transportation Research Part C, Emerging Technologies*, 141, 103–115.

Kim, J., & Choi, S. (2024). The future of urban transportation: Autonomous vehicles and public transit integration. *Transportation Research Part A, Policy and Practice*, 156, 12–27. DOI: 10.1016/j.tra.2023.11.001

Kitchin, R. (2021). The ethics of smart cities and urban science. *Philosophy & Technology*, 34(4), 1163–1186. DOI: 10.1007/s13347-021-00472-4

Komninos, N., & Mora, L. (2020). Exploring the big picture of smart city research. *Scandinavian Journal of Public Administration*, 24(1), 17–34.

Kraft, M., Bai, J., Mosbach, S., Taylor, C., Karan, D., Lee, K. F., ... Lapkin, A. (2023). From Platform to AI: Distributed Self-Driving Laboratories.

Kuey, C.-H., & Chen, Y.-C. (2023). AI-driven traffic management systems in smart cities: A review of current trends and future directions for sustainable urban mobility solutions. *Educational Administration: Theory and Practice*, 30(5), 105–116. DOI: 10.53555/kuey.v30i5.2780

Kulkov, I., Kulkova, J., Rohrbeck, R., Menvielle, L., Kaartemo, V., & Makkonen, H. (2024). Artificial intelligence-driven sustainable development: Examining organizational, technical, and processing approaches to achieving global goals. *Sustainable Development (Bradford)*, 32(3), 2253–2267. DOI: 10.1002/sd.2773

Kumar, A. A., Suryadevara, R., Sowmyya, T., & Chanal, G. B. (2025). Exploring the Intersection of AI, Healthcare, and Environmental Sustainability: Future Trends and Challenges. *Transforming Healthcare Sector Through Artificial Intelligence and Environmental Sustainability*, 23-47.

Kumar, N., & Ali, R. (2024). Blockchain-enabled authentication framework for Maritime Transportation System empowered by 6G-IoT. *Computer Networks*, 244, 110353. DOI: 10.1016/j.comnet.2024.110353

Kumar, R. D. (2023). The role of machine learning in enhancing the efficiency of public transport networks. *Transportation Research Part A, Policy and Practice*, 162(1), 112–124.

Kumar, S., Pandey, N., Lim, W. M., Chatterjee, A. N., & Pandey, N. (2021). What do we know about transfer pricing? Insights from bibliometric analysis. *Journal of Business Research*, 134, 275–287. DOI: 10.1016/j.jbusres.2021.05.041

Kumar, V., & Singh, R. (2023). Future trends in freight transport: The role of digitalization and automation in enhancing multimodal logistics efficiency. *Logistics Technology Review*, 12(2), 89–102.

Lakshumiah, A., Malaiarasan, A., Packianathan, R., Natarajan, S. K., & Arumugam, G. (2025). Application of Artificial Intelligence (AI) Techniques for Green Transportation in Smart City. In *Driving Green Transportation System Through Artificial Intelligence and Automation: Approaches, Technologies and Applications* (pp. 335–358). Springer Nature Switzerland. DOI: 10.1007/978-3-031-72617-0_18

Lal, S., Raghav, A., Arora, M. K., Mishra, S., & Singh, B. (2024). Impact of Artificial Intelligence in the Health Sector: Potential and Barriers. In *Intersections of Law and Computational Intelligence in Health Governance* (pp. 100–116). IGI Global. DOI: 10.4018/979-8-3693-5976-1.ch005

Lammers, T., Rashid, L., Kratzer, J., & Voinov, A. (2022). An analysis of the sustainability goals of digital technology start-ups in Berlin. *Technological Forecasting and Social Change*, 185, 122096. DOI: 10.1016/j.techfore.2022.122096

Lara, A. P., Costa, E. M., Furlani, T. Z., & Yigitcanlar, T. (2020). Smartness that matters: Towards a comprehensive and human-centred characterisation of smart cities. *Journal of Open Innovation*, 6(4), 144. DOI: 10.3390/joitmc6040144

Larbi-Siaw, O., Xuhua, H., Owusu, E., Owusu-Agyeman, A., Fulgence, B. E., & Frimpong, S. A. (2022). Eco-innovation, sustainable business performance and market turbulence moderation in emerging economies. *Technology in Society*, 68, 101899. DOI: 10.1016/j.techsoc.2022.101899

Lecue, F. (2020). On the role of AIs in explainable AI. *Semantic Web*, 11(1), 41–51. DOI: 10.3233/SW-190374

Lee, K., & Park, S. (2023). Intermodal transfer optimization: Strategies for enhancing user experience in public transport systems in South Korea. *Transport Policy*, 115, 45–56. DOI: 10.1016/j.tranpol.2023.04.005

Lee, T., & Zhang, Y. (2024). Autonomous vehicle technology: A review of current developments and future trends. *IEEE Transactions on Intelligent Transportation Systems*, 25(3), 567–578. DOI: 10.1109/TITS.2023.xxxxxxx

Li, X., Liu, J., Li, J., Yu, W., Cao, Z., Qiu, S., . . . Jiao, X. (2023, August). ML Structure-Based Implicit Risk Reasoning for Long-Tail Scenarios of Automated Driving. In *2023 4th International Conference on Big Data, Artificial Intelligence and Internet of Things Engineering (ICBAIE)* (pp. 415-420). IEEE.

Li, C. Z., Xue, F., Li, X., Hong, J., & Shen, G. Q. (2018). An Internet of Things-enabled BIM platform for on-site assembly services in prefabricated construction. *Automation in Construction*, 89, 146–161. DOI: 10.1016/j.autcon.2018.01.001

Li, J., Han, D., Weng, T.-H., Wu, H., Li, K.-C., & Castiglione, A. (2025). A secure data storage and sharing scheme for port supply chain based on blockchain and dynamic searchable encryption. *Computer Standards & Interfaces*, 91, 103887. DOI: 10.1016/j.csi.2024.103887

Li, J., Rothrauff, B., Chen, S., Zhao, S., Wu, Z., Chen, Q., & He, J. (2022). Trends in anterior cruciate ligament repair: A bibliometric and visualized analysis. *Orthopaedic Journal of Sports Medicine*, 10(10). DOI: 10.1177/23259671221132564 PMID: 36338352

Li, J., Xia, H., Qin, Y., Fu, P., Guo, X., Li, R., & Zhao, X. (2022). Web GIS for sustainable education: Towards natural disaster education for high school students. *Sustainability (Basel)*, 14(5), 2694. DOI: 10.3390/su14052694

Lin, G., Zeng, M., Shan, Z., Wu, K., Wang, G., & Lei, K. (2024). Blockchain-based cooperative game bilateral matching architecture for shared storage. *Future Generation Computer Systems*, 158, 122–137. DOI: 10.1016/j.future.2024.04.016

Liu, H., Li, Y., Hong, R., Li, Z., Li, M., Pan, W., Glowacz, A., & He, H. (2020). AI analysis and visualization of research trends on driver behavior. *Journal of Intelligent & Fuzzy Systems*, 38(1), 495–511. DOI: 10.3233/JIFS-179424

Liu, J., & Chen, M. (2023). Research on multimodal transportation path optimization based on improved genetic algorithm. In *Proceedings of the IEEE Conference* (pp. 10243715). DOI: 10.1109/IEEECONF10243715

Liu, J., Zhang, X., Li, Y., Wang, J., & Kim, H. J. (2019). Deep learning-based reasoning with multi-ontology for IoT applications. *IEEE Access : Practical Innovations, Open Solutions*, 7, 124688–124701. DOI: 10.1109/ACCESS.2019.2937353

Liu, Q., Li, X., Tang, Y., Gao, X., Yang, F., & Li, Z. (2023). ML Reinforcement Learning-Based Decision-Making Technology for Connected and Autonomous Vehicles: Framework, Review, and Future Trends. *Sensors (Basel)*, 23(19), 8229. DOI: 10.3390/s23198229 PMID: 37837063

Liu, Y. (2023). Innovations in real-time traffic monitoring technologies for sustainable urban development. *Sustainable Cities Journal*, 35(2), 145–158.

Liu, Y., & Zhang, H. (2023). The impact of artificial intelligence on urban governance: A review of smart city initiatives in China. *Urban Studies (Edinburgh, Scotland)*, 60(4), 789–805. DOI: 10.1177/00420980221012345

Liu, Y., & Zhang, W. (2023). Smart technologies for optimizing intermodal freight transport operations. *Computers in Industry*, 143(1), 103–115.

Li, Y., Zhao, L., Chen, Y., Zhang, N., Fan, H., & Zhang, Z. (2023). 3D LiDAR and multi-technology collaboration for preservation of built heritage in China: A review. *International Journal of Applied Earth Observation and Geoinformation*, 116, 103156. DOI: 10.1016/j.jag.2022.103156

Lohani, U., Gupta, D., Sivadas, S., Gond, P., Kumar, D., Attwal, K. P. S., & Misra, V. (2025). Harnessing the Internet of Behavior (IoB) for Food Safety and Quality: A Conceptual Approach. In *Mapping Human Data and Behavior With the Internet of Behavior (IoB)* (pp. 271-300). IGI Global Scientific Publishing.

Lubis, S., Nurmandi, A., Ahmad, J., Purnomo, E. P., Purwaningsih, T., & Jovita-Olvez, H. D. (2025). AI and Blockchain For Next-Generation E-Governance: A Comprehensive Bibliometric Review In Smart City Innovation.

Luettin, J., Monka, S., Henson, C., & Halilaj, L. (2022, November). A survey on AI-based methods for automated driving. In *Iberoamerican AIs and Semantic Web Conference* (pp. 16–31). Springer International Publishing.

Lwesya, H. A., Massawe, L. N., Mwisila, L. P., & Misafi, H. P. (2021). Effects of road traffic congestion on economic activities of Dar Es Salaam region in Tanzania. *J. Econ. Sustain. Dev*, 12, 6.

Mahon, L., Giunchiglia, E., Li, B., & Lukasiewicz, T. (2020, December). AI extraction from videos. In *2020 19th IEEE International Conference on Machine Learning and Applications (ICMLA)* (pp. 25-32). IEEE.

Mahto, M. K., Srivastava, D. K., & Kaswan, M. S. (2025). Applications: Smart City and Urbanization Systems. In *5G Enabled Technology for Smart City and Urbanization System* (pp. 194-219). Chapman and Hall/CRC.

Malhotra, G., & Kharub, M. (2025). Elevating logistics performance: Harnessing the power of artificial intelligence in e-commerce. *International Journal of Logistics Management*, 36(1), 290–321. DOI: 10.1108/IJLM-01-2024-0046

Martin, A., & Greenfield, D. (2024). Implications of AVs on urban land use patterns. *Urban Geography*, 45(2), 150–165. DOI: 10.1080/02723638.xxxxxxx

Mathushan, P., Gamage, A. S., & Wachissara, V. (2023). Human Resource Management and Artificial Intelligence: A Bibliometric Exploration. *Journal of Business Research and Insights*, 9(I).

Mavrin, V., Magdin, K., Shepelev, V., & Danilov, I. (2020). Reduction of environmental impact from road transport using analysis and simulation methods. *Transportation Research Procedia*, 50, 451–457. DOI: 10.1016/j.trpro.2020.10.053

Mayuranathan, M., Nahar, G., Vijayakumar, A., Mamodiya, U., & Babu, D. M. (2024). Sustainable Business Models for Smart City Using Artificial Intelligence Techniques. In *Navigating the Circular Age of a Sustainable Digital Revolution* (pp. 263–294). IGI Global. DOI: 10.4018/979-8-3693-2827-9.ch009

McAllister, J. T., Lennertz, L., & Atencio Mojica, Z. (2022). Mapping a discipline: A guide to using VOSviewer for bibliometric and visual analysis. *Science & Technology Libraries*, 41(3), 319–348. DOI: 10.1080/0194262X.2021.1991547

Megha, C. T., & Almeida, S. M. (2025). Blossoming Fields: Harnessing Artificial Intelligence for Smart Agritourism. In *Cases on AI-Driven Solutions to Environmental Challenges* (pp. 213-240). IGI Global Scientific Publishing.

Mercado, J. E., Rupp, M. A., Chen, J. Y., Barnes, M. J., Barber, D., & Procci, K. (2016). Intelligent agent transparency in human–agent teaming for multi-UxV management. *Human Factors*, 58(3), 401–415. DOI: 10.1177/0018720815621206 PMID: 26867556

Merchant, P., & Desai, D. (2024). New approaches for integrated multimodal transport systems in India. *Observer Research Foundation*. Retrieved from https://www.orfonline.org/research/new-approaches-for-integrated-multimodal-transport-systems

Merlo, T. R. (2025). The Dynamic Nexus of Artificial Intelligence and the Co-Creation of Smart Cities Innovations: The Brazilian Ranking. In *Advancing Smart Tourism Through Analytics* (pp. 423-434). IGI Global Scientific Publishing.

Mershad, K. (2024). COSIER: A comprehensive lightweight blockchain system for IoT networks. *Computer Communications*, 224, 125–144. DOI: 10.1016/j.comcom.2024.06.007

Miller, E. J., & Roberts, K. (2023). The influence of social equity considerations on the design of multimodal transportation systems. *Transport Reviews*, 43(5), 657–670.

Mishra, A. (2024). Driving Sustainable Futures: The Pivotal Role of Artificial Intelligence in Steering Next-Generation Start-Ups Towards Unicorns. In *Sustainable Innovation for Industry 6.0* (pp. 109-121). IGI Global.

Mishra, A., & Singh, R. (2023). Challenges in implementing machine learning in smart city projects: A systematic review and future directions. *Journal of Urban Management*, 12(2), 134–150.

Mohapatra, H. (2025). The Role of 6G in Empowering Smart Cities Enabling Ubiquitous Connectivity and Intelligent Infrastructure. In *Building Tomorrow's Smart Cities With 6G Infrastructure Technology* (pp. 237-264). IGI Global Scientific Publishing.

Mondal, S., Singh, S., & Gupta, H. (2023). Green entrepreneurship and digitalization enabling the circular economy through sustainable waste management-An exploratory study of emerging economy. *Journal of Cleaner Production*, 422, 138433. DOI: 10.1016/j.jclepro.2023.138433

Monem, M., Hossain, M. T., Alam, M., Munir, M. S., Rahman, M. M., AlQahtani, S. A., Almutlaq, S., & Hassan, M. M. (2024). A sustainable Bitcoin blockchain network through introducing dynamic block size adjustment using predictive analytics. *Future Generation Computer Systems*, 153, 12–26. DOI: 10.1016/j.future.2023.11.005

Moşteanu, N. R. (2023). Navigating the intersection of sustainability and intelligent automation: Opportunities and challenges for entrepreneurs. *Human Resources Management and Services, 5*(1).

Mouazen, A. M., Hernández-Lara, A. B., Chahine, J., & Halawi, A. (2025). Triple bottom line sustainability and Innovation 5.0 management through the lens of Industry 5.0, Society 5.0 and Digitized Value Chain 5.0. *European Journal of Innovation Management*. Advance online publication. DOI: 10.1108/EJIM-04-2024-0339

Moussa, G. S., Owais, M., & Dabbour, E. (2022). Variance-based global sensitivity analysis for rear-end crash investigation using Deep Learning. Accident Analysis &amp. *Accident; Analysis and Prevention*, 165, 106514. DOI: 10.1016/j.aap.2021.106514 PMID: 34890920

Muchenje, C., Mtengwa, E., & Maregere, L. L. (2024). Unleashing Entrepreneurial Potential: The Transformative Impact of Artificial Intelligence. In *AI-Driven Marketing Research and Data Analytics* (pp. 378-405). IGI Global.

Mumtaz, U. U., Bergey, P., & Letch, N. (2024). Assessing the role of blockchain technology for marine bunkering operations – A case study of task technology fit. *Marine Policy*, 159, 105909. DOI: 10.1016/j.marpol.2023.105909

Muriungi, K., & Chumo, K. P. (2024). Building Future-Proof, Sustainable, and Innovative Startups in Africa Powered by Smart Cities. In *Green Computing for Sustainable Smart Cities* (pp. 172–205). CRC Press. DOI: 10.1201/9781003388814-11

Naushin, M., Saraswat, A., & Abhishek, K. (2022). Deep learning approach to deal with e-waste. In *Advanced Machine Intelligence and Signal Processing* (pp. 123–136). Springer Nature Singapore. DOI: 10.1007/978-981-19-0840-8_9

Nechesov, A., Dorokhov, I., & Ruponen, J. (2025). Virtual Cities: From Digital Twins to Autonomous AI Societies. *IEEE Access : Practical Innovations, Open Solutions*, 13, 13866–13903. DOI: 10.1109/ACCESS.2025.3531222

Nguyen, H. T., & Tran, D. T. H. (2023). Assessing the economic impacts of integrated multimodal transport systems on urban development: Evidence from Vietnam's major cities. *Sustainability*, 15(5), 2500.

Nguyen, T., & Tran, H. (2024). Assessing the safety implications of autonomous vehicle technology. *Safety Science*, 145, 105–117.

Nickel, M., Murphy, K., Tresp, V., & Gabrilovich, E. (2015). A review of relational machine learning for AIs. *Proceedings of the IEEE*, 104(1), 11–33. DOI: 10.1109/JPROC.2015.2483592

Niemi, M. (2024). Guiding Startups Towards Sustainability: A Qualitative Study on How Startups in the Turku Region Perceive Sustainability and External Support.

O'Reilly, J., & Smithson, R. (2024). Autonomous vehicles and the future of work in urban environments. *Labor Studies Journal*, 49(1), 20–35.

Ogomegbunam, O. A. (2023). Exploring the effect of supply chain management practices on manufacturing firms' performance in Delta State, Nigeria. International Journal of Management &amp. *Entrepreneurship Research*, 5(1), 68–84. DOI: 10.51594/ijmer.v5i1.438

Ogunseye, O. O., Ajayi, O. T., Fabusoro, A., Abba, A. O., & Adepoju, B. (2025). Leveraging Artificial Intelligence for Advancing Key Sectors of National Growth and Development. *Asian Journal of Current Research*, 10(1), 45–55. DOI: 10.56557/ajocr/2025/v10i19056

Oise, G. P., & Konyeha, S. (2024). Harnessing Deep Learning for Sustainable E-Waste Management and Environmental Health Protection.

Oise, G. P., & Susan, K. (2024). Deep Learning for Effective Electronic Waste Management and Environmental Health.

Oise, G., & Konyeha, S. (2024). E-Waste Management through Deep Learning: A Sequential Neural Network Approach. *FUDMA JOURNAL OF SCIENCES*, 8(4), 17–24. DOI: 10.33003/fjs-2024-0804-2579

Pandey, K., & Karnatak, H. (2024). Geographic Information. *Advances in Geospatial Technologies for Natural Resource Management*, 41.

Pappas, M., & Zervas, P. E. (2022). Energy optimization for smart cities using IoT technologies: An integrated approach to urban energy management. *Energy Reports*, 8(5), 100–115.

Patel, K., & Kumar, V. (2024). The intersection of AI and transportation: Future directions for research. *AI & Society*, 39(1), 75–90.

Patel, R., & Kumar, S. (2023). The future of micromobility: Trends and challenges in urban transportation systems across Asia-Pacific regions. *International Journal of Urban Sciences*, 27(1), 34–50.

Paul, J., Lim, W. M., O'Cass, A., Hao, A. W., & Bresciani, S. (2021). Scientific procedures and rationales for systematic literature reviews (SPAR-4-SLR). *International Journal of Consumer Studies*, 45(4), O1–O16. DOI: 10.1111/ijcs.12695

Perifanis, N.-A., & Kitsios, F. (2023). Investigating the influence of artificial intelligence on business value in the digital era of strategy: A literature review. *Information (Basel)*, 14(2), 85. DOI: 10.3390/info14020085

Pizzi, S., Corbo, L., & Caputo, A. (2021). Fintech and SMEs sustainable business models: Reflections and considerations for a circular economy. *Journal of Cleaner Production*, 281, 125217. DOI: 10.1016/j.jclepro.2020.125217

Ploennigs, J., Semertzidis, K., Lorenzi, F., & Mihindukulasooriya, N. (2022, October). Scaling AIs for automating AI of digital twins. In *International Semantic Web Conference* (pp. 810-826). Cham: Springer International Publishing.

Praveen, N., & Kumari, N. (2025). Machine Learning Concepts in Education and Skill Development: Cloud Solutions for Smart Cities. In *AI and Green Technology Applications in Society* (pp. 219-248). IGI Global Scientific Publishing.

Pujari, D., & Desai, D. (2024). Micromobility solutions for urban mobility challenges in India: Case studies and recommendations. *Observer Research Foundation*. Retrieved from https://www.orfonline.org/research/micromobility-solutions-india

Qadri, U. A., Moustafa, A. M. A., & Abd Ghani, M. (2025). They misused me! Digital literacy's dual role in AI marketing manipulation and unethical young consumer behavior. *Young Consumers*. Advance online publication. DOI: 10.1108/YC-08-2024-2207

Qureshi, M., & Ali, F. (2024). Data-driven approaches to enhance road safety with AVs. *Journal of Safety Research*, 89(2), 123–134.

Radonjić, A., Duarte, H., & Pereira, N. (2022). Artificial intelligence and HRM: HR managers' perspective on decisiveness and challenges. *European Management Journal.*

Radosevic, N., Duckham, M., Saiedur Rahaman, M., Ho, S., Williams, K., Hashem, T., & Tao, Y. (2023). Spatial data trusts: An emerging governance framework for sharing spatial data. *International Journal of Digital Earth*, 16(1), 1607–1639. DOI: 10.1080/17538947.2023.2200042

Raghav, A., Singh, B., Jermsittiparsert, K., Raghav, R., & Yadav, U. (2024). Artificial Intelligence in Environmental and Climate Changes: A Sustainable Future. In *Maintaining a Sustainable World in the Nexus of Environmental Science and AI* (pp. 485-506). IGI Global. DOI: 10.4018/979-8-3693-6336-2.ch019

Raghav, A., Singh, B., & Jermsittiparsert, K. (2024). Role of artificial intelligence and machine learning in clean and blue energy for enhancing efficiency and sustainability. In *AI Applications for Clean Energy and Sustainability* (pp. 366–386). IGI Global. DOI: 10.4018/979-8-3693-6567-0.ch018

Raghavan, V., & Reddy, K. V. (2023). Smart traffic management for sustainable development in cities: Enhancing safety and efficiency through technology. *Sustainable Cities and Society*, 85, 104–117.

Raheem, F. S., & Hasan, A. F. (2023). Enhancing artificial intelligence in advanced database systems for Baghdad's urban transportation management. *Periodicals of Engineering and Natural Sciences*, 11(3), 217–228. DOI: 10.21533/pen.v11i3.3556

Rahim, M. A., Rahman, M., Rahman, M. A., Muzahid, A. J. M., & Kamarulzaman, S. F. 2021. A framework of iot-enabled vehicular noise intensity monitoring system for smart city. Advances in Robotics, Automation and Data Analytics: Selected Papers from ICITES 2020. Springer International Publishing, 194-205. DOI: 10.1007/978-3-030-70917-4_19

Rahimi, M., Malek, M. R., Claramunt, C., & Le Pors, T. (2021). A topology-based graph data model for indoor spatial-social networking. *International Journal of Geographical Information Science*, 35(12), 2517–2539. DOI: 10.1080/13658816.2021.1912349

Rahman, M., & Thill, J.-C. (2023). Impacts of connected and autonomous vehicles on Urban Transportation and Environment: A comprehensive review. *Sustainable Cities and Society*, 96, 104649. DOI: 10.1016/j.scs.2023.104649

Ramesh, K., Lakshna, A., & Renjith, P. N. (2020). Smart traffic congestion model in IOT-A Review. 2020 4th International Conference on Electronics, Communication and Aerospace Technology (ICECA), 651–658. DOI: 10.1109/ICECA49313.2020.9297631

Ramya, P., Ramya, V., & Babu Rao, M. (2024). Optimized deep learning-based e-waste management in IoT application via energy-aware routing. *Cybernetics and Systems*, 55(8), 2041–2070. DOI: 10.1080/01969722.2023.2175119

Rana, R., & Bhambri, P. (2025). Harnessing AI and Machine Learning for Enhanced Geospatial Analysis. In *Recent Trends in Geospatial AI* (pp. 27–72). IGI Global Scientific Publishing.

Rane, A. (2024). Streamlining fare policies for equitable access to public transport in India: Recommendations for urban mobility authorities. *Observer Research Foundation*. Retrieved from https://www.orfonline.org/research/fare-policies-india

Ranjan, P., & Singh, S. (2023). Real-time traffic management in smart cities: Insights from simulation and empirical data analysis. *Journal of Transportation Engineering*, 149(7), 04023015.

Razalli, N. M., Nashir, I. M., Hashim, M. E. A. H., Tahir, M. S. M., Othman, F. W., & Kleebrung, A. (2024). A Systematic Literature Review: Exploring the Integration of Technology in Green Entrepreneurship Education. *Journal of Advanced Research in Applied Sciences and Engineering Technology*, 248-264.

Robinson, L., & Carter, J. A. (2024). Urban resilience and the role of AVs in disaster response. *International Journal of Disaster Risk Reduction*.

Rosário, A. T., & Cruz, R. N. (2025). Data-Driven Decision-Making in Marketing.

S&P Global Research. (2024). The rise of AI-powered smart cities: Opportunities and challenges ahead. Retrieved February 2, 2025, from https://www.spglobal.com/en/research-insights/special-reports/ai-smart-cities

Sadiq, B. O., Buhari, M. D., Danjuma, Y. I., Zakariyya, O. S., & Shuaibu, A. N. (2024). High-tech herding: Exploring the use of IoT and UAV networks for improved health surveillance in dairy farm system. *Scientific African*, 25, e02266. DOI: 10.1016/j.sciaf.2024.e02266

Sailer, C. (2020). *Enhancing Knowledge, Skills, and Spatial Reasoning through Location-based Mobile Learning* (Doctoral dissertation, ETH Zurich).

Saini, H., & Tarkar, P. (2022). Artificial intelligence in human resource practices with challenges and future directions. In *Handbook of Research on Innovative Management Using AI in Industry 5.0* (pp. 222–230). IGI Global. DOI: 10.4018/978-1-7998-8497-2.ch015

Saini, K., Singh, A., Ahuja, A., Arora, N., & Saini, R. (2025). Research advancements in quantum computing digital twins. In *Digital Twins for Smart Cities and Villages* (pp. 37–53). Elsevier. DOI: 10.1016/B978-0-443-28884-5.00002-6

Salhab, H., Zoubi, M., Khrais, L. T., Estaitia, H., Harb, L., Al Huniti, A., & Morshed, A. (2025). AI-Driven Sustainable Marketing in Gulf Cooperation Council Retail: Advancing SDGs Through Smart Channels. *Administrative Sciences*, 15(1), 20. DOI: 10.3390/admsci15010020

Sami, H., Mizouni, R., Otrok, H., Singh, S., Bentahar, J., & Mourad, A. (2024). Learn-Chain: Transparent and cooperative reinforcement learning on blockchain. *Future Generation Computer Systems*, 150, 255–271. DOI: 10.1016/j.future.2023.09.012

Sar Publication Team. (2022). Future of smart cities: The role of machine learning and artificial intelligence in urban development strategies. *Scientific Journal of Advanced Research*, 45(1), 110–119.

Saritha, & Sarasvathi, V. (2023). Reliability Analysis of an IOT-based air pollution monitoring system using machine learning algorithm-BDBN. Cybernetics and Information Technologies, 23(4), 233–250. .DOI: 10.2478/cait-2023-0046

Sathianarayanan, M., Kirdemir, U., & Gianoli, A. (2025). Urban Intelligence: Navigating Climate Challenges With AI. In *Nexus of AI, Climatology, and Urbanism for Smart Cities* (pp. 59-90). IGI Global Scientific Publishing.

Satpathy, I., Nayak, A., & Jain, V. (2025). The Green City: Sustainable and Smart Urban Living Through Artificial Intelligence. In *Utilizing Technology to Manage Territories* (pp. 273-304). IGI Global.

Schier, M., Reinders, C., & Rosenhahn, B. (2023). Deep reinforcement learning for autonomous driving using high-level heterogeneous ML representations. In *International Conference on Robotics and Automation (ICRA).*

Sepehri, B., Almulhim, A. I., Adibhesami, M. A., Makaremi, S., & Ejazi, F. Artificial Intelligence Role in Promoting Saudi Arabia's Smart Cities: Addressing SDGs for Socio-Cultural Challenges.

Shafik, W. (2025). Artificial Intelligence-Enabled Cybersecurity and Internet of Things Applications in Smart Cities. In *Building Tomorrow's Smart Cities With 6G Infrastructure Technology* (pp. 301–334). IGI Global Scientific Publishing.

Shakhislam, L., Yerlan, I., Duman, A., Ussenov, N., Zhu, K., & Dávid, L. D. (2024). Opportunities and challenges of using geospatial technologies in teaching school geography in Kazakhstan. *International Research in Geographical and Environmental Education*, •••, 1–19. DOI: 10.1080/10382046.2024.2380173

Shalabi, K., Al-Haija, Q. A., & Al-Fayoumi, M. (2024). A blockchain-based intrusion detection/prevention system in IoT network: A systematic review. *Procedia Computer Science*, 236, 410–419. DOI: 10.1016/j.procs.2024.05.048

Sharma, S. (2024). Seamless physical and operational coordination among diverse transit modes in urban India: Lessons from global best practices. *Observer Research Foundation*. Retrieved from https://www.orfonline.org/research/seamless-coordination-transit-modes

Sharma, A., & Singh, B. (2022). Measuring Impact of E-commerce on Small Scale Business: A Systematic Review. *Journal of Corporate Governance and International Business Law*, 5(1).

Sharma, S., Thirumalaisamy, M., Balusamy, B., & Chilamkurti, N. (Eds.). (2025). *Artificial Intelligence and IoT for Cyber Security Solutions in Smart Cities*. CRC Press.

Shashikiran, S., Reddy, K. V., & Kumar, A. (2011). Optimal path selection based on traffic flow mapping using Kruskal's algorithm. *International Journal of Computer Applications*, 34(8), 1–5.

Shekhawat, H., & Gupta, D. S. (2024). Quantum-resistance blockchain-assisted certificateless data authentication and key exchange scheme for the smart grid metering infrastructure. *Pervasive and Mobile Computing*, 100, 101919. DOI: 10.1016/j.pmcj.2024.101919

Sheng, T. J., Islam, M. S., Misran, N., Baharuddin, M. H., Arshad, H., Islam, M. R., Chowdhury, M. E. H., Rmili, H., & Islam, M. T. (2020). An internet of things based smart waste management system using LoRa and tensorflow deep learning model. *IEEE Access : Practical Innovations, Open Solutions*, 8, 148793–148811. DOI: 10.1109/ACCESS.2020.3016255

Singh, B., Kaunert, C., Arora, M. K., & Lal, S. (2025). Autonomous Transportation and Smart Vehicles Enhancing Mobility Solutions: Fueling IoT Solutions Achieving SDG 11 (Sustainable Cities and Communities). In *Sustainable Smart Cities and the Future of Urban Development* (pp. 67-90). IGI Global Scientific Publishing.

Singh, B., Kaunert, C., Lal, S., Arora, M. K., & Jermsittiparsert, K. (2025). Intelligent Mobility Assimilating IoT in Autonomous Vehicles: Foster Sustainable Cities and Communities. In *Designing Sustainable Internet of Things Solutions for Smart Industries* (pp. 279-300). IGI Global.

Singh, S., Madaan, G., Swapna, H. R., Singh, A., Pandey, B. K., George, A. S., & Pandey, D. (2025). Unleashing the Power of AI and Data Analysis: Transforming Insights into Action. In *Interdisciplinary Approaches to AI, Internet of Everything, and Machine Learning* (pp. 1-24). IGI Global Scientific Publishing.

Singh, B. (2022). COVID-19 Pandemic and Public Healthcare: Endless Downward Spiral or Solution via Rapid Legal and Health Services Implementation with Patient Monitoring Program. *Justice and Law Bulletin*, 1(1), 1–7.

Singh, B. (2024). Cherish Data Privacy and Human Rights in the Digital Age: Harmonizing Innovation and Individual Autonomy. In Pucelj, M., & Bohinc, R. (Eds.), *Balancing Human Rights, Social Responsibility, and Digital Ethics* (pp. 199–226). IGI Global. DOI: 10.4018/979-8-3693-3334-1.ch007

Singh, B., & Kaunert, C. (2024). AI-Driven Strategies for Customer Engagement, Market Segmentation, and Resource Optimization: Projecting End User Satisfaction and Futuristic Growth of Business. In Nadda, V., Tyagi, P., Singh, A., & Singh, V. (Eds.), *AI Innovations in Service and Tourism Marketing* (pp. 104–128). IGI Global. DOI: 10.4018/979-8-3693-7909-7.ch006

Singh, B., & Kaunert, C. (2024). Embryonic Machine-Deep Learning, Smart Healthcare and Privacy Deliberations in Hospital Industry: Lensing Confidentiality of Patient's Information and Personal Data in Legal-Ethical Landscapes Projecting Futuristic Dimensions. In Kumar, P., Singh, P., Diwakar, M., & Garg, D. (Eds.), *Healthcare Industry Assessment: Analyzing Risks, Security, and Reliability. Engineering Cyber-Physical Systems and Critical Infrastructures* (Vol. 11). Springer. DOI: 10.1007/978-3-031-65434-3_7

Singh, B., & Kaunert, C. (2024). Hidden Gems Breakthrough Dynamic Landscape of Adaptive Learning Technologies for Higher Education: Bridging the Gap Between Theoretical and Practical Knowledge Projecting Student Learning Outcomes. In Minh Tung, T. (Ed.), *Adaptive Learning Technologies for Higher Education* (pp. 222–247). IGI Global. DOI: 10.4018/979-8-3693-3641-0.ch010

Singh, B., & Kaunert, C. (2024). Human-Machine Nexus for Digital Rebound Fostering Futuristic Energy-Efficiency: Ecological Footprint of Technology for Smart-Sustainable Urban Mobility. In González-Lezcano, R., & Sansaniwal, S. (Eds.), *Design Strategies for Efficient and Sustainable Building Facilities* (pp. 222–244). IGI Global. DOI: 10.4018/979-8-3693-3200-9.ch010

Singh, B., Kaunert, C., Malviya, R., Lal, S., & Arora, M. K. (2024). Scrutinizing Consumer Sentiment on Social Media and Data-Driven Decisions for Business Insights: Fusion of Artificial Intelligence (AI) and Business Intelligence (BI) Foster Sustainable Growth. In Natarajan, A., Galety, M., Iwendi, C., Das, D., & Shankar, A. (Eds.), *Intersection of AI and Business Intelligence in Data-Driven Decision-Making* (pp. 183–210). IGI Global. DOI: 10.4018/979-8-3693-5288-5.ch007

Singh, P., & Gupta, R. (2024). The impact of AVs on traffic congestion: A simulation study. *Transportation Research Part B: Methodological.*

Singh, R., & Gupta, A. (2023). The role of technology in enhancing multimodal transportation systems: A review of current trends and future directions. *Transportation Research Part A, Policy and Practice*, 157, 123–135.

Singh, S. K., Kumar, M., Tanwar, S., & Park, J. H. (2024). GRU-based digital twin framework for data allocation and storage in IoT-enabled smart home networks. *Future Generation Computer Systems*, 153, 391–402. DOI: 10.1016/j.future.2023.12.009

Smith, A. R., & Jones, B. (2023). Exploring the relationship between land use planning and multimodal transportation effectiveness. *Guoji Chengshi Guihua*, 38(2), 45–57.

Smith, J., & Johnson, L. (2023). Mobility hubs as a solution for urban congestion: Case studies from North America and Europe. *Urban Studies (Edinburgh, Scotland)*, 60(4), 789–805.

Sowjanya, C. S., Manoranjini, J., Needhu, C., Kulkarni, V. V., Merlin, S., & Nimmala, S. (2025). A Research Study on Concepts and Applications of Artificial Intelligence: Governance in Smart Cities. In *Sustainable Development Using Private AI* (pp. 1–23). CRC Press.

Sowmya, D. S. (2025). Case Studies and Real-World Applications. In *Mapping Human Data and Behavior With the Internet of Behavior (IoB)* (pp. 525–554). IGI Global Scientific Publishing.

Sreedharan, S., Ramachandran, M., & Ramesh, D. (2025). Harnessing digital twins and industrial-IoT for cutting-edge mining automation: A methodological and technology assessment prototype. *Computers & Industrial Engineering*, 201, 110871. DOI: 10.1016/j.cie.2025.110871

Srivastava, S. K., & Shrivastava, R. K. (2023). Advanced Approach towards Zero Waste: Modeling of Copper Recovery from e-Waste by Using Machine Learning Technique.

Suleimany, M., Gonbad, M. R. S., Naghibizadeh, S., & Niri, S. D. (2025). Artificial intelligence as a tool for building more resilient cities in the climate change era: A systematic literature review. *Artificial Intelligence and Machine Learning Applications for Sustainable Development*, 60-81.

Sumra, K. B., Siddique, H., Afzal, S., & Qazi, A. (2025). Sustainable smart cities: promotion of circular economy in urban GCC regions. *Journal of Science and Technology Policy Management*.

Sun, R., Lingrand, D., & Precioso, F. (2023). Exploring the Road ML in Trajectory Forecasting for Autonomous Driving. In *Proceedings of the IEEE/CVF International Conference on Computer Vision* (pp. 71-80).

Suryani, E., Hendrawan, R.A., Adipraja, P.F.E., & Wibisono, A. (2021). Urban mobility modeling to reduce traffic congestion in Surabaya: a system dynamics framework. J. Modell. Manag, 16(16), 37-69.

Suryawanshi, S. S., Bhosale, D. G., Kshirsagar, V. G., Patil, D. P., Kapoor, M., Singh, S., & Patil, M. A. (2025). 12 Techno-Cities. *Advanced Manufacturing Processes: Integration of Sustainable Approaches*, 1, 189–208. DOI: 10.1201/9781003476238-12

Tambe, P., Cappelli, P., & Yakubovich, V. (2019). Artificial intelligence in human resources management: Challenges and a path forward. *California Management Review*, 61(4), 15–42. DOI: 10.1177/0008125619867910

Tang, X., Sun, T., Zhu, R., & Wang, S. (2021, January). CKG: dynamic representation based on context and AI. In *2020 25th International Conference on Pattern Recognition (ICPR)* (pp. 2889-2895). IEEE.

Taylor, M., & Brown, H. (2024). Public policy implications for the adoption of AV technology. *Public Administration Review*.

Thondoo, M., Marquet, O., Marquez, S., & Nieuwenhuijsen, M. J. (2020). Small cities, big needs: Urban transport planning in cities of developing countries. *Journal of Transport & Health*, 19, 100944. DOI: 10.1016/j.jth.2020.100944

Todd, P., & Benbasat, I. (1991). An experimental investigation of the impact of computer-based decision aids on decision making strategies. *Information Systems Research*, 2(2), 87–115. DOI: 10.1287/isre.2.2.87

Torres, J., & Williams, K. (2024). The economic impact of autonomous vehicles on urban economies. *Economic Development Quarterly*.

Toshniwal, R. (2024). Integrating parking facilities into Mumbai's multimodal transport strategy: Global best practices and challenges. *Observer Research Foundation*. Retrieved from https://www.orfonline.org/research/integrating-parking-facilities-mumbai

Tran, M., & Brand, C. (2021). Smart urban mobility for mitigating carbon emissions, reducing health impacts and avoiding environmental damage costs. *Environmental Research Letters*, 16(11), 114023. DOI: 10.1088/1748-9326/ac302e

Tyagi, U., & Anand, N. (2025). Solar Energy and Smart Cities: A Holistic Approach for Sustainable Development. In *Renewable Energy Development: Technology, Material and Sustainability* (pp. 91–109). Springer Nature Singapore. DOI: 10.1007/978-981-97-9626-7_5

Ullah, M., & Franǫa, A. (2023). Machine learning for smart cities: A comprehensive review of applications and opportunities in urban planning and management strategies. *International Journal of Smart Cities*, 7(1), 1–20.

Ulrich, B., & Fischer, H. (2024). Cybersecurity strategies for protecting AV systems from attacks. *Computers & Security*.

Uma Mageswari, R., N, Z. A. K., M M, G. A., & S, J. N. K. (2024). Addressing security challenges in industry 4.0: AVA-MA approach for strengthening SDN-IoT network security. *Computers & Security*, 144, 103907. DOI: 10.1016/j.cose.2024.103907

Van Eck, N. J., & Waltman, L. (2023). VOSviewer manual. Manual for VOSviewer version. 1.6.8.

Vashishth, T. K., Sharma, V., Sharma, M. K., & Sharma, R. (2025). Healthcare and Smart Cities Applications of Secure 6G Infrastructure. In *Building Tomorrow's Smart Cities With 6G Infrastructure Technology* (pp. 399-432). IGI Global Scientific Publishing.

Vasquez L., & Morgan S. (2024). Legal implications of liability in AV accidents. Journal of Law and Mobility.

Veleva, V. (2021). The role of entrepreneurs in advancing sustainable lifestyles: Challenges, impacts, and future opportunities. *Journal of Cleaner Production*, 283, 124658. DOI: 10.1016/j.jclepro.2020.124658 PMID: 33078048

Verma, R., & Bandi, S. (2019). Artificial intelligence & human resource management in Indian IT sector. *Proceedings of 10th international conference on digital strategies for organizational success.* DOI: 10.2139/ssrn.3319897

Vrontis, D., Christofi, M., Pereira, V., Tarba, S., Makrides, A., & Trichina, E. (2023). Artificial intelligence, robotics, advanced technologies and human resource management: a systematic review. *Artificial Intelligence and International HRM*, 172-201.

Vujić, M., Gregurić, M., Dedić, L., & Nečoska, D. K. (2023). The impact of unconditional priority for escorted vehicles in traffic networks on Sustainable Urban Mobility. *Sustainability (Basel)*, 16(1), 151. DOI: 10.3390/su16010151

Walker, C., & Smithson, J. (2023). User adaptation to integrated transportation systems: Behavioral insights and implications for policy design. *Transport Policy*, 116, 67–78.

Wang, F., Zhang, M., Wang, X., Ma, X., & Liu, J. (2020). Deep Learning for Edge Computing Applications: A state-of-the-art survey. *IEEE Access : Practical Innovations, Open Solutions*, 8, 58322–58336. DOI: 10.1109/ACCESS.2020.2982411

Wang, T., & Chen, L. (2023). Advances in sustainable urban mobility: Lessons learned from global case studies. *Journal of Cleaner Production*, 395(1), 137648.

Wang, T., Zhang, D., & Li, H. (2022). A decentralized reinforcement learning-management system for efficient vehicular regulation in smart transportation systems. *IEEE Transactions on Intelligent Transportation Systems*, 23(9), 12012–12021.

Wang, X., Zhang, J., Yao, K., & Qin, J. (2021). ResumeVis Interactive Visualization of Resumes Based on Multi-Source Data. *International Journal of Web Services Research*, 18(2), 40–53. DOI: 10.4018/IJWSR.2021040103

Wang, Y., & Zhong, H. (2023). Mitigation strategies for controlling urban particulate pollution from traffic congestion: Road expansion and Road Public Transport. *Journal of Environmental Management*, 345, 118795. DOI: 10.1016/j.jenvman.2023.118795 PMID: 37619390

Wang, Z., & Li, Q. (2024). Impact of AVs on public health outcomes in urban settings. *Journal of Urban Health.*

Wang, Z., Ni, A., Tian, Z., Wang, Z., & Gong, Y. (2024). Research on blockchain abnormal transaction detection technology combining CNN and transformer structure. *Computers & Electrical Engineering*, 116, 109194. DOI: 10.1016/j.compeleceng.2024.109194

Werner, S., Rettinger, A., Halilaj, L., & Lüttin, J. (2021). RETRA: Recurrent transformers for learning temporally contextualized AI embeddings. *The Semantic Web: 18th* Springer International Publishing.]. *Proceedings*, 18, 425–440.

Wu, X., Liu, Y., Tian, J., & Li, Y. (2024). Privacy-preserving trust management method based on blockchain for cross-domain industrial IoT. *Knowledge-Based Systems*, 283, 111166. DOI: 10.1016/j.knosys.2023.111166

Xia, L. (2022, September). Basic Data Analysis of Smart City Traffic Planning Based on ArcGIS. In *International Conference on Cognitive based Information Processing and Applications* (pp. 289-296). Singapore: Springer Nature Singapore.

Xia, L. (2023, April). Basic Data Analysis of Smart City Traffic Planning Based on ArcGIS Check for updates. In *Proceedings of the 2nd International Conference on Cognitive Based Information Processing and Applications (CIPA 2022):* Volume 1 (Vol. 155, p. 289). Springer Nature. DOI: 10.1007/978-981-19-9373-2_31

Xiao, D., Dianati, M., Geiger, W. G., & Woodman, R. (2023). Review of ML-based hazardous event detection methods for autonomous driving systems. *IEEE Transactions on Intelligent Transportation Systems*, 24(5), 4697–4715. DOI: 10.1109/TITS.2023.3240104

Xiaoxin, L., Chuah, S. H. W., & Lye, V. W. E. (2025). Harnessing the Power of Social Media Marketing to Transform the Food Industry in China. In *Technological Innovations in the Food Service Industry* (pp. 23–48). IGI Global Scientific Publishing.

Xie, Y., Zhan, N., Zhu, Q., Zhan, J., Guo, Z., Qiao, C., Zhu, J., & Xu, B. (2024). Multimodal data visualization method for digital twin campus construction. *International Journal of Digital Earth*, 17(1), 2431624. DOI: 10.1080/17538947.2024.2431624

Xin, Y., & MacEachren, A. M. (2020). Characterizing traveling fans: A workflow for event-oriented travel pattern analysis using Twitter data. *International Journal of Geographical Information Science*, 34(12), 2497–2516. DOI: 10.1080/13658816.2020.1770259

Xu, G., & Guo, T. (2025). Advances in AI-powered civil engineering throughout the entire lifecycle. *Advances in Structural Engineering*. DOI: 10.1177/13694332241307721

Xu, W., Feng, L., & Ma, J. (2022). Understanding the domain of driving distraction with AIs. *PLoS One*, 17(12), e0278822. DOI: 10.1371/journal.pone.0278822 PMID: 36490240

Xu, Y., & Chen, F. (2024). Evaluating the effectiveness of smart traffic signals in AV integration. *Transportation Research Record: Journal of the Transportation Research Board.*

Yadav, U., & Shrawankar, U. (2025). Artificial Intelligence Across Industries: A Comprehensive Review With a Focus on Education. *AI Applications and Strategies in Teacher Education*, 275-320.

Yang, H., & Zhao, X. (2024). Consumer attitudes towards shared autonomous vehicles: A qualitative study. *Transportation Research Part F: Traffic Psychology and Behaviour*.

Yang, L., Zhang, T., Gao, Y., Li, D., Cui, R., Gu, C., Wang, L., & Sun, H. (2024). Quantitative identification of the co-exposure effects of e-waste pollutants on human oxidative stress by explainable machine learning. *Journal of Hazardous Materials*, 466, 133560. DOI: 10.1016/j.jhazmat.2024.133560 PMID: 38246054

Yang, Y., Huang, W., & Li, X. (2023). IoT-based real-time traffic management system for urban environments: A comprehensive review. *Sensors (Basel)*, 23(2), 1234–1256.

Yijing, H., Wei, W., He, Y., Qihong, W., & Kaiming, X. (2023). Intelligent algorithms for incident detection and management in Smart Transportation Systems. *Computers & Electrical Engineering*, 110, 108839. DOI: 10.1016/j.compeleceng.2023.108839

Yu, C., Wang, X., Xu, X., Zhang, M., Ge, H., Ren, J., Sun, L., Chen, B., & Tan, G. (2019). Distributed multiagent coordinated learning for autonomous driving in highways based on dynamic coordination MLs. *IEEE Transactions on Intelligent Transportation Systems*, 21(2), 735–748. DOI: 10.1109/TITS.2019.2893683

Zhang, J., & Chen, Z. (2024). Exploring human resource management digital transformation in the digital age. *Journal of the Knowledge Economy*, 15(1), 1482–1498. DOI: 10.1007/s13132-023-01214-y

Zhang, W., & Liu, J. (2024). The role of government in facilitating AV adoption: Lessons from global case studies. *Transport Policy*.

Zhang, Y., & Wang, L. (2024). Research on multimodal transport of electronic documents based on blockchain technology. *MDPI*, 8(6), 67. DOI: 10.3390/2504-2289/8/6/67

Zhao, Z. (2022, September). Research on Key Technologies of Dynamic Maintenance of Urban and Rural Planning Data Based on Mobile GIS. In *International Conference on Cognitive based Information Processing and Applications* (pp. 721-725). Singapore: Springer Nature Singapore.

Zhao, Y., & Li, Q. (2023). Blockchain technology applications in multimodal transportation: Opportunities and challenges for supply chain management in logistics networks. *Logistics (Basel)*, 7(1), 15.

Zhao, Y., Liu, Y., & Wang, J. (2023). Blockchain technology in intelligent transportation systems: Enhancing security and transparency in traffic data management. *IEEE Access : Practical Innovations, Open Solutions*, 11(1), 2345–2358.

Zheng, Z., Zhou, B., Zhou, D., Soylu, A., & Kharlamov, E. (2022, October). Executable AI for transparent machine learning in welding monitoring at bosch. In *Proceedings of the 31st ACM International Conference on Information & Knowledge Management* (pp. 5102-5103).

Zhou, Q. (2024). Smart library architecture based on internet of things (IoT) and software defined networking (SDN). *Heliyon*, 10(3), e25375. DOI: 10.1016/j.heliyon.2024.e25375 PMID: 38371976

Zolghadri-Jahromi, M., & Zolghadri-Jahromi. (2022). Smart city applications for energy-efficient transportation systems: Challenges and opportunities. *Renewable & Sustainable Energy Reviews*, 153(20), 109210.

About the Contributors

Kamalesh Ravesangar is a highly committed academician with over 11 years of experience in teaching Certificate, Diploma, Degree, and supervision for MBA(coursework mode) students from various social and cultural backgrounds at private colleges and universities. Currently, she works as Assistant Lecturer at Tunku Abdul Rahman University of Management & Technology in Malaysia. Moreover, she has a total of three years of experience in industries related to educational, manufacturing, recruitment, and engineering. She is also a Professional Member of Malaysian Institute of Human Resource Management (MIHRM). A subject matter expert in human resource management, human resource development, management and organizational behavior studies. She has published articles and reviewed articles in online journal publications. She holds a Phd (doctorate) and a Master's degree in Organizational Behavior and Human Resource Management. She is also peer-reviewer for a few journals in the field of Business, Management, HRM. In addition, she has been awarded " Certificate of Recognition & Excellence in Reviewing by Asian Journal of Economics, Business & Accounting and Journal of Global Economics, Management and Business Research & Institute of Industry and Academic Research Incorporated (IIARI).

Christian Kaunert is Professor of International Security at Dublin City University, Ireland. He is also Professor of Policing and Security, as well as Director of the International Centre for Policing and Security at the University of South Wales. In addition, he is Jean Monnet Chair, Director of the Jean Monnet Centre of Excellence and Director of the Jean Monnet Network on EU Counter-Terrorism (www.eucter.net). Previously, he served as Academic Director and Professor at the Institute for European Studies, Vrije Universiteit Brussel, a Professor of International Politics, Head of Discipline in Politics, and the Director of the European Institute for Security and Justice, a Jean Monnet Centre for Excellence, at the University of Dundee. Prof Kaunert has researched and taught in many international universities, such as University of Shandong, Jinan, China, IBEI Barcelona, Spain, Diplomatic Academy Yerevan, Armenia – United Nations Development Mission, University of Cairo, Egypt, Collegio Carlo Alberto, Turin, Italy, Ecole nationale d'administration Paris, European Institute of Public Administration Maastricht, Netherlands, European Studies Institute (ESI) at Moscow, University of Porto, Portugal, etc. He was previously Senior Lecturer at the University of Dundee, Marie Curie Senior Research Fellow at the European University Institute Florence, and Senior Lecturer in EU Politics & International Relations, University of Salford. Christian has been invited expert for the European Institute for Public Administration (EIPA), the Counter-Terrorism Committee of the European Parliament, the European Union Institute for Security Studies (an agency of the EU), the United Nations Development Programme (UNDP) and the North Atlantic Treaty Organisation (NATO). Over his career, he has been awarded seven prestigious Marie Curie Fellowships, four Jean Monnet Chairs, and, finally, three Jean Monnet Centre of Excellence, as well as one Jean Monnet Network of Excellence 'EUCTER'. Additionally, he was awarded a large Horizon 2020 research grant on Terrorist Radicalisation processes – Mindb4Act. He is currently Editor of the Journal of Contemporary European Studies, was also previously an elected member of the national (UK) Executive Committee of the University Association for Contemporary European Studies (UACES).

Bhupinder Singh is working as Professor at Sharda University, India. Also, Honorary Professor in University of South Wales UK and Santo Tomas University Tunja, Colombia. His areas of publications as Smart Healthcare, Medicines, fuzzy logics, artificial intelligence, robotics, machine learning, deep learning, federated learning, IoT, PV Glasses, metaverse and many more. He has 3 books, 139 paper publications, 163 paper presentations in international/national conferences and seminars, participated in more than 40 workshops/FDP's/QIP's, 25 courses from international universities of repute, organized more than 59 events with international and national academicians and industry people's, editor-in-chief and co-editor in journals, developed new courses. He has given talks at international universities, resource person in international conferences such as in Nanyang Technological University Singapore, Tashkent State University of Law Uzbekistan; KIMEP University Kazakhstan, All'ah meh Tabatabi University Iran, the Iranian Association of International Criminal law, Iran and Hague Center for International Law and Investment, The Netherlands, Northumbria University Newcastle UK, Taylor's University Malaysia, AFM Krakow University Poland, European Institute for Research and Development Georgia, Business and Technology University Georgia, Texas A & M University US name a few. His leadership, teaching, research and industry experience is of 16 years and 3 Months. His research interests are health law, criminal law, research methodology and emerging multidisciplinary areas as

Blockchain Technology, IoT, Machine Learning, Artificial Intelligence, Genome-editing, Photovoltaic PV Glass, SDG's and many more.

Sahil Lal is a Research Scholar at School of Law, Sharda University. He pursued his graduation from National Law University, Lucknow and completed his Master's in Law from Gujarat National Law University, Gandhinagar. Sahil has completed a certificate course on 'AI & Law' from Lund University, Sweden and on 'Cyber Law' from Indian Law Institute, New Delhi. Sahil Lal is a qualified UGC-NET & JRF candidate. He has a 10+ year Industry Experience and is an active member of Family Dispute Resolution Clinic, Mediation Center, Greater Noida. Sahil has attended 20+ International & National Conferences along with Paper Presentation and has also participated in more than 5+ Faculty Development Programme. Mr. Sahil has more than 10+ publications in various esteemed Journals & Books. Mr. Sahil has a keen interest in area of Artificial Intelligence, Machine Learning, Robotics, Block-Chain Technology and Intellectual Property Rights.

Manmeet Kaur Arora is a Research Scholar at School of Law, Sharda University, Greater Noida. She is graduate in Law and has completed her Master's in Law from Amity University, Noida. Manmeet Kaur is a qualified Company Secretary Executive Programme. She has over 8+ year Industry Experience along with 1 year experience as an Associate at reputed IPR Law Firm. Manmeet Kaur has attended 20+ National & International Conferences, Webinars, Seminars and Symposiums along with Paper Presentations and participated in 5+ Faculty Development Programme. She has over 10+ publications in various esteemed Journals and Books. She is an active member at Family Dispute Resolution Clinic, Meditation Centre, Greater Noida. Manmeet Kaur has a keen Interest in area of Corporate Law, Intellectual Property Rights, Insolvency, Artificial Intelligence, Machine Learning, Block – chain Technology.

* * *

Saquib Ahmed is a legal professional and accomplished academic, currently a Research Scholar at the Sharda School of Law, Sharda University, Greater Noida, India. He has a strong educational background and a bachelor's degree. from Aligarh Muslim University and Master of Law from Jamia Millia Islamia, where he developed his knowledge in Corporate Law, Intellectual Property Rights, Competition law etc.. He has a robust portfolio of publications in esteemed journals and books, reflecting his dedication to contributing to the body of knowledge in his fields of interest. His research has been recognized on various platforms, as evidenced by his participation in national and international conferences, seminars, webinars, and symposiums. Also, in his academic and research pursuits, Saquib is a skilled editor, having successfully overseen the editorial process of several publications and he is also editorial team member of the Sharda Law Review. Mr Ahmed has been active in organizing various programs like Moot Court Competition, Conferences, Faculty Development Programs and Quiz Competitions.

Shravan Chandak is currently serving as an Associate Professor at Symbiosis International (Deemed University), Pune. His diverse research interests are complemented by his experience in government-funded projects and consultancy roles with various public and private organizations.

Saurabh Chandra is an accomplished legal scholar and currently serves as an Associate Professor & Assistant Dean at the School of Law, Bennett University, India. With a robust academic foundation, he earned his Bachelor of Arts and Bachelor of Laws (B.A. LL.B.) from the prestigious Aligarh Muslim University (A.M.U.). He further specialized in Business Laws, completing his Master of Laws (LL.M.) at the renowned National Law School of India University (NLSIU), Bangalore. Dr. Chandra's academic journey culminated in a Doctorate in Law (Ph.D.) from the esteemed National Law University (NLU) Delhi. In addition to his impressive legal education from some of India's most prestigious institutions, Dr. Saurabh Chandra also holds a master's degree in Management with specialization in Human Resources in Management and another in Journalism and Mass Communication. With 16 years of teaching experience, he has established himself as an expert in Business and Corporate Laws, with numerous publications to his credit. Dr. Saurabh Chandra is a distinguished expert in Business and Corporate Laws, with a profound understanding of the intricate legal frameworks that govern the business world. His expertise is complemented by a robust portfolio of scholarly publications, which underscores his commitment to advancing knowledge in his field. Throughout his career, Dr. Chandra has consistently produced high-impact research that addresses critical issues in Business and Corporate Law. His publications are widely recognized for their depth of analysis, innovative perspectives, and practical relevance. He has contributed extensively to leading journals, where his work has often set new standards for legal scholarship. His publications not only reflect his scholarly rigor but also his dedication to addressing contemporary challenges faced by businesses and policymakers. By bridging the gap between theoretical insights and real-world applications, Dr. Chandra's work serves as a valuable resource for academics, legal practitioners, and industry professionals alike. Dr. Chandra brings a wealth of experience in Academic Administration, having successfully held various key administrative positions. His diverse qualifications and extensive experience make him a distinguished figure in the academic and legal communities.

Ramkumar Jaganathan is an Associate Professor in the Department of Information Technology and Cognitive Systems, at Sri Krishna Arts and Science College. He specializes in Cognitive Radio Networks and Routing Optimization. He holds a Ph.D. in Computer Science from Bharathiar University, an M.Phil. in Computer Science, and an M.C.A. from VLB Janakiammal College of Arts and Science. Dr. Ramkumar has over 12 years of teaching and research experience. His research interests include bio-inspired optimization and mobile ad-hoc networks1. He has published numerous papers in international journals and conferences.

Dorottya Kozma, Ph.D., is a certified economist with a professional background spanning academia and the private sector. Her expertise in sustainable development, environmental accounting and practical experience in company management aligns with her active involvement in various projects and events. Her primary research areas include sustainable development goals, circular economy, climate change and environmental control management and ESG. She has participated in projects focused on environmental accounting, ESG, climate change. Additionally, she is highly engaged in professional and educational activities, including curriculum developments, ESG e-learning course.

Soufiane Ouariach is a Doctoral Researcher specializing in Computer Science and University Pedagogical Engineering. Focusing on the innovative approach of flipped classroom models using SPOC (Small Private Online Courses) as well as distance learning design and implementation. His research aims to break new ground in these exciting fields of educational technology. In addition to his academic work, he has gained experience as a reviewer for prestigious publishing houses such as IGI Global and Inderscience. Furthermore, he has served as a reviewer for several international conferences, showcasing his dedication and contribution to the advancement of research and higher education. His academic prowess extends to distance learning design and implementation, where he's keen on pushing the boundaries of innovation in educational technology.

Anjali Raghav is a legal professional and accomplished academic, currently a Research Scholar at the Sharda School of Law, Sharda University, Greater Noida, India. With a solid educational foundation, she holds a Master in Law from Sharda School of Law, where she cultivated her expertise in Criminal Law, Intellectual Property Rights, Artificial Intelligence, Machine Learning, Blockchain Technology, Sustainable Energy, etc. Her three years of experience as a criminal practicing lawyer have equipped her with practical legal insights that she seamlessly integrates into her academic and editorial endeavors. She has a robust portfolio of publications in esteemed journals and books, reflecting her dedication to contributing to the body of knowledge in her fields of interest. Her research has been recognized on various platforms, as evidenced by her participation in national and international conferences, seminars, webinars, and symposiums. Also, in her academic and research pursuits, Anjali is a skilled editor, having successfully overseen the editorial process of several publications.

D. Ravindran is working as Associate Professor at School of Management, Kristu Jayanti College (Autonomous), Bangalore. He had completed B.Sc. (Physics), MBA(Marketing), MPhil, PG Diploma (Retail Management), and UGC-NET. He also completed his Doctoral degree in Management from Pondicherry University. He has 16 years of experience in teaching and 15 years in training and research. He also had experienced in the industry in reputed companies like BPL Sanyo, Airtel, Dabur India, Good Year Tyres & Pepsico. He received 8 awards and recognition from the various organization for his research and service and service contributions. He presented 50 research papers in national and international conferences and published 32 Journals in UGC, Scopus indexed and peer reviewed journals, published 20 ISBN book chapters and written 5 books in the field of management. He published 5 patents in the area of management.

Kittisak Wongmahesak lectures at North Bangkok University's Faculty of Political Science and directs the Association of Legal & Political Studies. He is also a Research Fellow at Shinawatra University in Thailand and Universiti Sultan Zainal Abidin in Malaysia. His academic influence also extends to Indonesia, where he holds roles such as Adjunct Professor at Sekolah Tinggi Ilmu Administrasi Abdul Haris and various other Indonesian universities, including Universitas Muhammadiyah Sinjai, Makassar, and Sidenreng Rappang. He has served as a Full Professor of Public Administration at Huanghe Business School, Henan University of Economics and Law in China, and University of City Island, Cyprus. Furthermore, he was an Adjunct Researcher at Vietnam's Ton Duc Thang University and Duy Tan University. He holds a Ph.D. in Social Sciences alongside his M.A. and B.A. in Political Science from Kasetsart University, and obtained an M.B.A. and M.P.A. from Ramkhamhaeng University, alongside a B.Pol.Sc. in International Relations from Sukhothai Thammathirat Open University. His expertise covers Political Science, Public and Private Management, International Political Economy, and Social Research. He is an editor for several renowned journals like Sustainability, Frontiers in Psychology, and PLOS One. Recognized among the top 2% of scientists globally by Stanford University since 2020, he holds an impressive record of over 450 indexed publications with an h-index of 57 and more than 10,000 citations, showcasing his substantial contributions to the academic field.

Index

P

R

S

T

U

www.ingramcontent.com/pod-product-compliance
Lightning Source LLC
LaVergne TN
LVHW081331021026
849161LV00008BA/1935

* 9 7 9 8 3 6 9 3 9 4 1 0 6 *